DATE		

THE ECONOMICS OF MINERAL EXTRACTION

THE ECONOMICS OF MINERAL EXTRACTION

- *Gerhard Anders*
- *W. Philip Gramm*
- *S. Charles Maurice*
- *Charles W. Smithson*

PRAEGER

PRAEGER SPECIAL STUDIES • PRAEGER SCIENTIFIC

Library of Congress Cataloging in Publication Data

Main entry under title:

The Economics of mineral extraction.

    Bibliography: p.
    Includes index.
    1. Mineral industries. I. Anders, Gerhard.
HD9506.A2E23   338.2   79-22949
ISBN 0-03-053171-3

Published in 1980 by Praeger Publishers
CBS Educational and Professional Publishing
A Division of CBS, Inc.
521 Fifth Avenue, New York, New York 10017 U.S.A.

123456789 038 98765432

Printed in the United States of America

Preface

The primary purpose of this book is to provide the reader with a pragmatic approach to the analysis of public policy issues in the mineral-extracting sector. To accomplish this purpose, the book is divided into two parts. In the first part (Chapter 2 through 5) the objective is to present a consistent, comprehensive analytical framework within which mineral-extracting firms and industries can be examined in general. In this part, the primary determinants of price, output, investment, etc. in the mineral-extracting sector will be identified and the reaction of the industries to changes in these determinants will be examined. In the second part of the book (Chapter 5 through 11), the objective is to provide a methodology for evaluating the impact of governmental policies on the mineral-extracting sector. Using the general analytical framework developed in the first part of the book, we consider the effect of alternative policies, such as taxation, conservation, and environmental controls on investment, output, and price, and provide some illustrative empirical measures of the actual *magnitudes* of the impact of specific policies.

The book is designed to be useful and interesting not only to persons trained in economics or to students of economics but also to persons directly involved in the industry itself or in the formulation of public policy, but who are not professional economists. We have therefore attempted to present the great majority of the material in such a way as to be easily understood by readers who have little or no formal training in economics. On the other hand, we have done our best to discuss the issues in a way that will not be trivial to professional economists. We hope and believe that most of the material will be of interest to both audiences.

Portions of certain chapters (for example, Chapter 2 and 4) may be a bit difficult for the reader who is not a trained economist to follow. We always indicate the portions that those readers may omit without loss of continuity or understanding of the remainder of the material. Moreover, we summarize the technical results in a nontechnical manner for any readers, noneconomists or economists, who do not choose to read or work through the technical material. This verbal summary can also be used as a synopsis for those who have in fact read the more technical material. Let us emphasize that we always indicate where this more technical material begins and summarize the results.

We should note that almost all the technical analysis that would prove difficult for the noneconomist is in the first part of the book: Chapters 2 through 5. The second part of the book, which is concerned basically with policy issues, should be easily followed by readers who are not economists and of general interest to readers who are trained in economics. Therefore, while we feel that the book forms a coherent whole, we also feel that it will be of use to readers interested only in policy issues. At the extreme, with only a reading of the summary of the basic theoretical analysis in Chapter 2, readers can, without loss of continuity, turn immediately to the chapters on the policy issues of interest in Part II.

Finally, we might note that there are a few portions of the book, primarily in Chapter 6, that may be rather trivial to professional economists, even though they may be of considerable interest to other readers. We always alert readers to these sections, which economists can easily omit, also.

This book represents essentially a consolidation and expansion of work carried out by the authors over the past five years. Most of this work was carried out under contract with or within the Ontario Ministry of Natural Resources. The authors gratefully acknowledge the support over this period of three consecutive Ministers of Natural Resources: The Hon. Leo Bernier, who supported the research at its inception; The Hon. Frank Miller, under whom the project gained momentum; and The Hon. James Auld, who allowed it to mature. Dr. J.K. Reynolds, the Deputy Minister of Natural Resources, and Mr. A.J. Herridge, the Assistant Deputy Minister, not only gave continued support but also showed great patience and understanding. Mr. G.A. Jewett, Executive Coordinator of the Mines Group, and Dr. T.P. Mohide, Director of the Mineral Resources Branch, supported the authors on a day-to-day basis and also helped greatly to keep the practical objectives in focus. The authors appreciate the ministry's permission to use material from three monographs written by the authors: *The Impact of Taxation and Environmental Controls on the Ontario Mining Industry; Investment Effects on the Mineral Industry of Tax and Environmental Policy Changes: A Simulation Model;* and *Factor Substitution and Biased Technical Change in the Canadian Mining Industry.*

Much of the empirical analysis contained in this book would have been far less precise if not impossible without the assistance of people in the mineral-extracting industry. We wish to thank particularly Mr. M.W. Airth of Noranda Mines, Mr. G.H.D. Hobbs of Cominco Ltd., and Drs. T. Podolsky and F. Gormley of Inco Ltd.

The authors would also like to thank the colleagues who provided support and assistance for this project. At Texas A&M University we wish to thank particularly Robert B. Ekelund, Roy F. Gilbert, Hae-Shin Hwang, and Thomas R. Saving. We also thank Yvonne Evans and Billie Ribbe for typing the manuscript and Lynn Gillette for checking mathematical computations.

Dr. William R. Allen of UCLA and the International Institute for Economic Research is acknowledged for his support and for permission to use material contained in *Does Resource Conservation Pay?* We also thank the

Richard D. Irwin Co. for permission to use portions of Chapter 11 of *Economic Analysis*, 3d edition, by C.E. Ferguson and S.C. Maurice.

Finally, we thank and acknowledge particularly the assistance of Gordon L. Bennett who, during and after the writing of his dissertation, worked at the ministry to obtain data, provided numerous suggestions and insights, and assisted in the analysis.

June 1979

Gerhard Anders
W. Philip Gramm
S. Charles Maurice
Charles W. Smithson

Contents

LIST OF TABLES

LIST OF FIGURES

THE ECONOMICS OF
MINERAL
EXTRACTION

1

Introduction

Concern about the future of natural resources, particularly the depletable mineral resources, became prominent in the early 1970s.* The world-wide economic boom had led to an extraordinary increase in the demand for resources and therefore to shortages and rapid price increases. It was during this same period that the "doomsday" or "limits to growth" theories were gaining wide public attention and acceptance. These theories, which in the extreme have generated predictions of a natural-resource-based Armageddon for the developed nations, were based on an underlying assumption that the prevailing trends in resource recovery and utilization would continue into the future.[1] They neglected, as have the Malthusian predictions of the past, the fact that existing trends may be altered by changes in the economic, technological, or political environment. It follows that, in order to evaluate this issue systematically, one must obtain information about potential future trends in the recovery and utilization of these resources. In this book, we will concentrate on the recovery aspect. We will show how market forces and governmental policies can affect the future rate of recovery (i.e., extraction) of mineral commodities.

---

*The "energy crisis" of the 1970s, however, was certainly not the first such crisis. An example of an earlier crisis from which we apparently learned little is the "whale oil crisis" of the nineteenth century, discussed in Chapter 10.

Closely related to the preceding issue is the concern about international resource cartels, exemplified best by OPEC. The ability of these cartels to transform economic power into political power has led the developed nations to reexamine their own commodity positions. While our analysis does not deal directly with international mineral commodity markets, we point out several instances in which internal factors have had a significant impact on a nation's dependence on foreign supplies. The effect of these internal factors is most clearly demonstrated in our discussion of the impact of taxation and environmental controls (Chapters 7 and 8) and of the percentage depletion allowance (Chapter 11).

Another general reason for considering mineral extraction is the growing concern about developing economies. Although the developed nations are the major exporters of many of the mineral commodities, minerals (including mineral fuels) represent approximately 44 percent of total exports for the developing economies.[2] It follows then that concerns such as equity in the international economy must take into consideration the workings of the extractive industries.

Within the purview of economics itself, the mineral-extracting sector merits specific attention because of the uniquely dynamic form of the production process. A primary feature of this industry is that, unlike most other producing sectors, production in any given period is not independent of production in any other period. Since the current rate of extraction affects the amount that may be extracted in future periods, the cost of extracting a unit of the mineral commodity today depends not only on the current level of usage of the inputs and their prices but also on the levels of input usage in the past and on the impact of current extraction on the future profitability of the deposit. Hence, in our analysis we assume that the firm attempts to maximize profit over its time horizon (i.e., its net worth) rather than maximizing profit in a specific time period. This distinction might at first seem artificial, since all firms attempt to maximize net worth. However, in the more traditional case a firm's attempt to maximize net worth may well imply maximizing profit in each period (so one can use a single-period model and consider the objective of the firm to be the maximization of profit in that period); this is clearly not the case in the mineral-extracting industries. Furthermore, current activity in mineral extraction can affect the level of reserves in a future period in two ways. While an increase in the current rate of extraction may reduce the level of reserves for a specific deposit, this increased activity can induce additional exploration and development, which would increase the level of reserves in a future period. It is with these dynamic features and some additional economic characteristics of the mineral-extracting firm and industry that the first part of this book (Chapters 2 through 5) will be concerned.

Finally, the mineral-extracting industries are of particular interest, since the sector is one that historically has been subject to considerable government intervention. While the motives behind the intervention have varied—including growth, self-sufficiency, environmental concerns, and conservation —such intervention has an impact on the levels of investment, input wages, output, and price in the mineral-extracting sector. In the second part of the book we explicitly consider several specific governmental policies. These include general taxation and environmental controls (Chapters 6 through 9), enforced conservation through stockpiling and production holdbacks (Chapter 10), and the percentage depletion allowance, a tax treatment specific to the extractive industries (Chapter 11).

While a considerable body of economic literature concerned with mineral-extracting firms and industries exists, these works are generally limited to a single topic or issue. In this book we will attempt to provide a more comprehensive view of the mineral-extracting sector. To this end, we first develop and examine the basic models of the individual mineral-extracting firm and the mineral-extracting industry in Part I. We then use these models to examine the impact of governmental policies in Part II. This use of a common theoretical approach will make it possible to see how the seemingly disjointed policy actions are related. Throughout the book, we will, when applicable, provide a brief summary of the preceding literature on which our analyses are based.

In our analyses, both of the mineral-extracting firm and industry in general and of the impact of governmental policies, we will provide empirical evidence to examine our hypotheses and to illustrate our techniques. For example, in the case of the mineral-extracting firm and industry in general we provide evidence concerning the competitiveness of the industries and some of the characteristics of the production relation in mining. In the case of the impact of governmental policies we employ historical data to estimate such things as the magnitudes of the impact of changes in taxation and environmental controls on investment and of the impact of the percentage depletion allowance on output and price. While we use a considerable amount of data from the United States, most of the data concerned with mining (other than for mineral fuels) are drawn from the mineral-extraction sector in Canada. This should not, however, limit the applicability of our results. Canada represents an excellent source of data because it not only is important in the international mineral commodity markets but also is more dependent on mining for the generation of wealth, employment, and export earnings than almost any other industrialized nation. Canada accounts for approximately 9 percent of the world's reserves of copper, 15 percent of iron ore, 13 percent of lead, 14 percent of nickel, and 26 percent of zinc.[3] In 1976 Canada's production of these minerals relative to total world production was 9 percent for copper, 6 percent for iron ore, 7 percent for lead, 31 percent for nickel and 16 percent for zinc.[4] Illustrative of its impact on the Canadian economy, mineral extraction (excluding mineral fuels) accounts for 3.4 percent of GNP, 0.8 percent of total employment, and 7.4 percent of total exports. In the United States, for instance, these figures would be 0.3 percent, 0.1 percent, and 1.5 percent, respectively.[5]

We have attempted to write this book so that it can be used not only by persons trained in economics but also by noneconomists directly involved in the industry itself or in the formulation of public policy. However, while we have tried to present the discussion in such a way that it does not rely heavily on either economic jargon or a substantial background in economics, some chapters or portions of chapters may prove to be difficult for the noneconomist. Such difficulties may arise in either the theoretical analysis (e.g., Chapters 2 and 5) or the empirical analysis (e.g., Chapter 4). In light of this problem, these areas difficult for the noneconomist are accompanied by a nontechnical summary that may be used as a synopsis or in lieu of the more technical analysis. Likewise, we have included some material specifically for the noneconomist that may be skipped by the reader trained in economics. Hence, the book can be of use to a wider range of readers. Indeed, for the reader interested in a specific topic, it is possible to read only the nontechnical summary of Chapter 2, which contains our basic theoretical models, and move immediately to the topic of particular interest. Let us now turn to a brief overview of the specific topics considered.

As noted above, the basic theoretical models used in the analyses of specific policies are developed in Chapter 2. We employ a competitive framework for the individual firm and, more importantly, for the mineral-extracting industry (this crucial feature is examined empirically in Chapter 3). We first show that, for the individual firm, the current rate of extraction affects the quantity that can be extracted in future periods; therefore, the cost of extracting an additional unit of the mineral commodity in the current period depends not only on the cost of the additional inputs that must be employed today but also on the level of input usage (i.e., the rate of extraction) in past periods and on the effect of current extraction on the future profitability of the deposit. Since the owner or leasor of the deposit has some time horizon, he will maximize total profits over that horizon—the net worth—rather than simply maximizing profit in any given period. The firm will select that rate of extraction and, since the entire mineral deposit is not generally exploited, the total level of extraction that maximizes the present value of the deposit. It follows directly from this that the firm's level of investment must also be such as to maximize the net worth of the deposit. A firm will undertake an investment so long as the marginal rate of return from the investment exceeds the cost of the investment (i.e., the interest rate). Anything having the effect of raising (or lowering) the marginal returns from investment in mining will induce firms to increase (or decrease) investment, thereby increasing (or decreasing) output. It is at the industry level that this feature is most important. Assume, for instance, that government policies are altered such that the marginal returns to investment in mining rise relative to those in the rest of the economy. The increase in marginal returns will cause additional investment to flow into mining, lowering the rate of return earned on the additional investment, until the rate of return earned in the mineral-extracting sector is again equal to that for the economy as a whole. Hence, the impact of governmental policies may be ascertained by their effect on the marginal return to investment.

In Chapter 3 the competitiveness of the mineral-extracting sector is examined empirically. As is demonstrated theoretically in Chapter 2, competition would drive the after-tax rate of return in mining to that obtainable in the economy as a whole. Using Canadian data, two ratios were calculated to indicate the rate of return: the ratio of profits after taxation to capital employed and the ratio of profits after taxation to total equity. The ratios for mining were compared to those for manufacturing during the period 1962–75. In neither case did the rate of return for mining differ significantly from that for manufacturing. Using data from industries traded on the financial markets in the United States, the rates of return (calculated as the ratio of profits after taxation and depreciation to book value) for manufacturing and mining were compared. Again, there was no significant difference. Hence, the available data are consistent with our assumption that the mining industry is competitive.

To a large extent, the ability of the mineral-extracting industries to react to changes in market conditions and/or governmental policy changes is determined by the underlying production technology. More specifically, the degree of factor substitutability and the bias (if any) in technical change can affect the relative response of the industries. In Chapter 4 the production relation is examined empirically using data drawn from Canadian mining industries over the period 1962–74.

A final issue of importance in the examination of the mineral-extracting firm and industry in general is the decision to adopt a new or nonconventional technology. Illustrations of such technologies are deep-seabed mining and the use of peacetime nuclear explosives. In Chapter 5 the basic models developed in Chapter 2 are expanded to provide a methodology that may be employed to examine the economic feasibility of the implementation of new technologies. The major difference between our analysis and those that have preceded it is that we consider economic rather than technological feasibility; the fundamentals of the demand for and supply of the mineral commodity are explicitly incorporated. Most of the preceding studies employed what we might call a "gapsmanship" approach. They assumed that, for some reason, quantity demanded exceeds quantity supplied and that this "gap" must be filled via nonconventional technology. They further assumed that the additional costs of the higher-priced technology are completely passed on to the consumer. The problem with such analyses is that they failed to consider the effect of price on quantity demanded. In contrast, our analysis indicates that, in a competitive market, nonconventional production would lower rather than raise resource prices. Costs have an effect on price, but so does demand. If quantity produced is to rise, price must fall, unless there exists an externally imposed shortage (e.g., a government-imposed ceiling price). We conclude this analysis with the outline of a methodology that could be employed to examine the economic feasibility of a specific nonconventional technology.

Chapter 6 begins a section in which the impact of general taxation and environmental controls will be examined. This introductory chapter aimed primarily at the noneconomist begins with a discussion of externalities—particularly pollution—in order to demonstrate the economic aspects of the problem. An externality is said to exist when the actions of one economic unit causes the cost conditions facing another unit to change. For instance, pollution is a negative externality since the actions of a polluting firm impose additional costs on persons residing in the area (i.e., some of the costs of operation are passed from the firm to the residents of the area). In contrast to some more market-oriented solutions to this situation that are discussed in this chapter, environmental controls operate by placing restrictions on the amount of pollution permitted. We then proceed to show that these environmental controls would impact on the sector in which they are imposed by reducing the marginal return on investment. It follows then that the economic effect would be a reduction in investment and therefore a reduction in output and a corresponding increase in price.

The effects of taxation are explicitly introduced in Chapter 7. As was the case with environmental controls, increased taxation leads to a reduction in the marginal return to investment and therefore to a reduction in investment and output. Given that the theoretical effects of environmental controls and taxation are well established, we turn to a method of estimating the *magnitude* of such effects. We provide some simple predictive models for the magnitude of the impact of environmental controls and taxation on the mineral-extracting industries. Using the basic models developed in Chapter 2, the reduction in investment—resulting from either the imposition of environmental controls or increases in taxation—relative to the level of investment that would have been forthcoming in the absence of such changes can be measured via the change in the present value of an investment. Since this change can itself be quantified using our model of the mining industry, estimates of the magnitude of the impact can be obtained.

Chapter 8 provides an application of the methodology developed in Chapter 7. We use the predictive model to evaluate the changes in taxation and environmental legislation that occurred in Canada in the early 1970s. Several alternative combinations of tax and environmental policies are presented in order to illustrate the individual and joint effects of these policy changes. More specifically, we provide simulations of what investment will be relative to what it would have been considering (1) the changes in environmental legislation only, (2) the changes in taxation only, and (3) the impact of the changes in both. Since some of the exogenous variables cannot be estimated with a high level of precision, each of these simulations was performed using a range of values (i.e., high, low, and "best" estimates) for these exogenous variables. The resulting estimates then provide a range of magnitudes for the impact of the changes in governmental policy. Nonetheless, our estimates indicate that the impact of these changes is quite substantial. Indeed our estimates indicate that investment in mining would be 30 to 65 percent less than what would have occurred in the absence of these changes, with our "best" estimates indicating a 43-percent reduction.

The examination of the effects of general taxation and environmental controls is concluded in Chapter 9. From Chapter 8 it is clear that the cost to society of the imposition of environmental controls can be quite substantial. The economic benefits from environmental controls would, therefore, have to be very large for the controls to be economically justified (i.e., in order that the benefits exceed the costs). Chapter 9 analyzes the benefits theoretically and examines them empirically. A crucial point of this chapter is that the benefits from controls are difficult to quantify and even more difficult to assign. For instance, if a person moved into an area that has substantial pollution, he may be compensated through lower housing prices. If environmental controls are imposed later, that individual would be additionally compensated (i.e., he will receive a "windfall" gain on the value of his property), while the person who left the area as a result of the pollution is not compensated for the loss he sustained on the value of his property. In the case of mining, this problem is compounded by the fact that many of the individuals damaged by the pollution are associated with the firm in other ways—they lease land to the firm, are employed by the firm, etc.—and may have already been compensated for the pollution through higher rents and wages. These difficulties notwithstanding, we do provide some measure of the economic benefits of environmental controls via a conceptual experiment. Using Canadian data, we calculate what the effect on GNP would have been had the existing environmental controls been instituted earlier and had all or some of the "disturbed" land been converted to agricultural or forestry uses. While we do not even suggest that these estimates are numerically exact, the point is that the purely economic benefits from controls fall far short of the costs.

The conservation issue is taken up in Chapter 10. In this discussion both stockpiling and the extraction holdbacks are considered as alternative methods of conservation. From the point of view of economics, such policies would be viable only if the discounted value of expected future returns exceed the returns obtainable from current production. An increase in the value of the resource over time alone is not sufficient. The increase in the expected value over time must exceed the returns that could be expected from selling the resource now and investing the income. Hence, the key variable in this decision is the rate of interest. While this is formulated in the context of a profit-maximizing firm, the analysis is relevant for government as well, since the sacrificed return reflects the true cost to the society of withholding resources. This issue was examined empirically using data for 14 depletable resources over the period 1900–75. In no case would stockpiling have been a viable economic alternative in the long run. Indeed, there is no evidence that resources were extracted too rapidly and it may have been that resources have been extracted at too slow a rate during the twentieth century. Two major conclusions, largely in contradiction with general opinion, result from the study. First, at any time during the twentieth century, enforced long-term conservation would have been a poor economic decision, not only for the generation giving up consumption but also for the future generations. Second, whatever motivations—military or political—may be used for stockpiling, the exhaustion of resources has historically not been a valid justification.

The final topic considered is the percentage depletion allowance, a tax treatment specific to extracting industries. In Chapter 11 the basic model developed in Chapter 2 is employed to provide a methodology for estimating the magnitude of the impact of percentage depletion on the output and the price of the mineral commodity. Using data from the U.S. petroleum industry for the period 1951–71, years of relative stability in the industry, we estimated what price and output would have been in absence of the percentage depletion allowance. Then we obtained an estimate of the magnitude of the impact by comparing these simulated values with the actual values. Our estimates indicate that the impact of depletion is significant. Indeed, even the lowest estimates would indicate that, if the depletion allowance had been eliminated in 1951, output in 1971 would have been 25 percent lower and price would have been 25 percent higher.

NOTES

1. While there are numerous books and articles of this type, the seminal work is Donella H. Meadows, Dennis L. Meadows, Jorgen Randers, and William W. Behrens III, *The Limits to Growth* (New York: Universe, 1972).

2. Rachel McCulloch and José Piñera, "Alternative Commodity Trade Regimes," in *Sharing Global Resources* (New York: McGraw-Hill, 1979).

3. Ruth W. Arad and Uzi B. Arad, "Scarce Natural Resources and Potential Conflict," in *ibid.*

4. Metallgesellschaft Aktiengellschaft, *Metal Statistics* (Frankfurt am Main: Metallgesellschaft Aktiengellschaft, 1978).

5. These figures were collected from various sources in "The Decline of Small Mineral Enterprises in Ontario," a discussion paper prepared by the Mineral Resources Group, Ontario Ministry of Natural Resources, 1978.

PART I

THE MINERAL-
EXTRACTING FIRM
AND INDUSTRY

2

Theory of
the Mineral-extracting Firm and
Industry

The purpose of this chapter is to lay the foundation for much of the analysis to follow. Here we set up the basic models or theories to be used in analyzing specific aspects of mineral-extracting industries.

We will set forth the basic theories rigorously, sometimes using mathematics that may be unfamiliar to the reader who is not a trained economist. But as we stressed in Chapter 1, this book is designed not only for professional economists but also as much, if not more so, for noneconomists who are interested in the economics of natural resources. Therefore, the results of the theories set forth are discussed in nontechnical language, easily understandable by people who are not economists. We shall always indicate those portions of the analysis that can be omitted by the noneconomist with no loss of continuity. Moreover, we include some explanations of economic terms and justification of the use of types of economic analysis that are quite well known and obvious to economists but not to noneconomists. We shall indicate these parts of the chapter that can be omitted by economists also.

Because this chapter is theoretical in nature, the conclusions that can be drawn from the models are quite general. The analysis is not designed to apply to any *specific* mineral firm. Let us emphasize, however, that the theories are designed to apply to "the real world of mineral extraction."

The use of theoretical analysis to analyze problems of the real world is generally acceptable to most economists, but we probably do need to justify this approach to the reader who is not an economist. One frequently hears statements such as, "That's okay in theory, but how about the real world?" In fact, a popular economist said recently on a TV show, "They [the planners of a particular country] don't go by theory, they go by what works." Belief in misleading statements such as these is extremely detrimental, particularly for decision makers in government or business.

Economic theory allows us to gain insights into the economy or segments of the economy that would be impossible without the use of a theoretical structure. Using simple theories, we can analyze the potential results of certain activities and make predictions that would be impossible without the use of a theoretical structure. Economic theory enables us to explain things about the real world even though those theories abstract from most actual attributes of the world.

To summarize, the purpose of theory is to make sense out of confusion. The real world, or the segment of it in which we are interested, is a complicated place in which an infinite number of variables are in continual change. Economic theory is designed to allow us to concentrate on the few forces that are important to the issue at hand and to ignore the infinite number of variables that are of little or no importance. In other words, when using a theory we try to abstract from the irrelevant, enabling us to keep from becoming bogged down in unimportant issues. When we attempt to measure or predict the impact of some external force, theory allows us to determine which data are important and which are not. Certainly, we shall use a considerable amount of statistics in our analysis, but these statistics will be used within a sound theoretical structure.

For example, if we wish to predict the effect of a change in an economy's tax structure on the amount of copper produced, there are relatively few pieces of data that would be of importance. Some of these would be the elasticity of demand for copper (how responsive quantity demanded is to price), the cost of producing copper, the returns to factors of production that extract copper, the amount of copper reserves, the structure of the copper market, the degree to which inputs may be substituted, and so on. But, as we all know, the price and output of copper depend to a small extent, possibly to an infinitesimally small extent, on a large number of other factors. Our theory allows us to abstract away from these unimportant influences.

Thus, we shall draw conclusions and make predictions using simple assumptions, while ignoring forces that *could* affect the results but in all likelihood will not. Our answers will be essentially correct, but in carrying out our analysis we must remember that while everything depends upon everything else, most things depend fundamentally upon only a *few* other things. If pressed far enough, the price of copper can be shown to depend not only on the cost of producing copper and the prices of substitute metals but also upon the prices of beef, airline tickets, and beer. But, in general, we ignore the effect of unimportant aspects of the problem at hand in order to concentrate on a few closely related variables.

This chapter sets forth several of the fundamental theories used to analyze problems in the remainder of this book. Some of the theoretical structure is based to a greater or lesser extent on the works of others, other parts are original. We will attempt primarily to set forth theories that will be useful in analyzing the mineral industry and will also be easily understood by noneconomists. In particular, the theoretical model that will be employed in this analysis recognizes the inherently dynamic nature of mineral extraction and incorporates this characteristic using traditional techniques.

We begin with an analysis of depletion production and cost functions and then use these functions to develop the theory of the mineral-extracting firm and industry. We shall explain how prices and output are determined. We next develop a theory of investment and exploration in mineral extraction. We then turn to a discussion of industry supply and demand and conclude the chapter with an examination of the implications of long-run equilibrium for a mineral-extracting industry in an individual economy. Since some of our theoretical structure employs the analyses of preceding researchers, a survey of the literature concerned with the mineral-extracting firm and industry is contained in the appendix to this chapter.

DEPLETION PRODUCTION AND COST FUNCTIONS

The cost of production or extraction of minerals is based upon two primary components: the underlying production function and the prices of inputs used to produce the mineral. We will begin with the production function. In the typical textbook case it is generally assumed that production or output in one period depends upon the state of the technology and the amounts of resources or inputs used in that period. While this method is quite useful in a large number of cases, a complicating feature is encountered in the case of extractive industries. The problem in mineral extraction is that the amount produced or extracted during a specific period affects the productivity of inputs in future periods.

Let us assume a somewhat simplified case. Suppose there is an ore deposit that contains only one grade of ore. One firm owns the entire deposit, and its time horizon extends from the present period to some period in the future. The firm recognizes that production or output in any specific period depends not only upon the amounts of each input used in that period but also upon the input usage in all preceding periods. This interdependency among periods results because the amount of extraction in previous periods affects the productivity of inputs in the current period.

This type of production relation leads to a cost function in which production in the current period increases the cost of production in future periods. For example, assume the amount extracted in period 1 is X tons. Then the cost of extracting X tons in period 2 will be Y dollars. But with the assumed production function, if less than X tons is extracted in period 1, the cost of extracting X tons in period 2 will be *less than* Y dollars. If more than X tons is extracted in period 1, the cost of extracting X tons in period 2 will be more than Y dollars. The same results hold for future periods.

Define marginal cost as the change in total cost per unit change in output. In the "usual" case marginal cost is the increase in cost during a particular period, per unit increase in output in that same period. But because of the peculiarities of extractive industries, the true marginal cost of extraction is *the increase in cost in the current period plus the increase in costs of extracting a given level of output in future periods due to extraction in the current period.* This type of cost function is the basic format used in our theories of mineral extracting firms.

We can express these relations more specifically for the mathematically inclined reader. The extraction production function in say the i-th period can be written as follows:*

$$q_i = f(x_1^i, x_2^i, \ldots, x_n^i, x_1^{i-1}, x_2^{i-1}, \ldots, x_n^0). \tag{2.1}$$

where q_i is output in the i-th period
          x_k^i is the amount of input k used in the i-th period
          x_r^{i-j} is the amount of the r-th input used in period $(i-j)$ for all
                $j = 0, 1, 2, \ldots, i-1.$

The marginal product is

$$\frac{\partial q_i}{\partial x_k^i} > 0.$$

The effect of past input usage on the amount of output possible in period i from a given vector of inputs used in the i-th period can be expressed as

$$\frac{\partial q_i}{\partial x_r^{i-j}} \leq 0.$$

for $(j = 0, 1, 2, \ldots, i-1)$

This term simply indicates that additional input usage in past periods can lower the possible output from a given vector of inputs in the current period

---

*A more extensive discussion of the specific form of the production function in this industry is taken up in Chapter 4.

because of the greater production in the past resulting from greater input usage in the past. Finally, the marginal product of a given input in the i-th period, expressed as $\partial q_i / \partial x_k^i$, can be decreased by additional usage of inputs in preceding periods.

This type of production function means that the more ore extracted now, the greater the cost of extracting a specific amount of ore in future periods.[1] Thus the cost function of extracting in period i can be written as

$$C_i = C_i(q_i, q_{i-1}, q_{i-2}, \ldots, q_0, Z) \tag{2.2}$$

where Z is the amount extracted prior to period zero (in the case in which the firm is not beginning extraction). For the type of cost function shown in equation (2.2) the *total* marginal cost of extraction in period i is

$$MC = \frac{\partial C_i}{\partial q_i} + \sum_{t=i+1}^{H} \frac{\partial C_t}{\partial q_i}, \tag{2.3}$$

where $\partial C_i / \partial q_i$ is the "typical" marginal cost of extraction in the current period and is positive. The term $\partial C_t / \partial q_i$ is the effect of extraction in period i on the cost of extraction in some future period t. Each of these terms is equal to or greater than zero for all t greater than i.

THEORY OF THE FIRM

Most of the literature concerning the mineral-extracting firm as described in the appendix to this chapter concerns two fundamental decisions that must be made: (1) the rate at which to extract and (2) the length of the period over which to extract. In much of the literature, particularly the earlier works, it was simply assumed that the firm has a fixed quantity of the mineral that it would extract over its time horizon. The firm merely decides the rate at which to extract; when the ore body is exhausted, the time horizon ends.

The Simplest Model

Let us assume first that the firm has a fixed quantity of a mineral and will extract until the total amount is extracted. The firm maximizes the present value or discounted stream of profit, written as

$$PV = \sum_{t=0}^{H} \left(\frac{1}{1+r}\right)^t (R_t - C_t); \tag{2.4}$$

subject to a fixed amount of the mineral resource, where

r   is the relevant rate of discount
H   is the end of the firm's time horizon
R_t is gross revenue in period t
C_t is total cost in period t

The constraint on the firm requires that the firm extracts a certain amount and no more. We will assume that it extracts no less.

We will also assume that the prices in each period are known by and given to the firm by the market. That is, the firm by itself has no effect on the price of the mineral. The choice variables for the firm are the amounts to be extracted in each period. Maximization of present value under these circumstances requires that the firm in each period extract the amount at which discounted price in that period equals discounted *total* marginal cost.

Before going further, an additional word or two about the makeup of marginal cost for an extractive firm such as that modeled here would be in order. If the firm had a deposit that was unlimited in the sense that it would not be exhausted during the time horizon of the firm, and if the rate of extraction during one period did not affect the cost of extraction in future periods, the marginal cost would simply be the rate at which current output increases current costs. As noted in the previous section, the model used here is more complicated. Moreover, the constraint that the firm has a fixed amount of the resource complicates the model even further.

Recall from above that marginal cost for an extractive industry is the effect of current output on current cost plus the discounted sum of the effects of current output in the current period on the costs in all future periods. But, with a fixed amount of the mineral resource there is an additional marginal cost generally called a "user cost." Thus, user cost shows the opportunity cost of current extraction on future profit, because a unit of output extracted now cannot be extracted in the future. Thus this "lost" future profit from the reduction in the quantity of the mineral available is as real a cost to the firm as the increase in present cost due to current extraction. These three components of marginal cost will be derived mathematically below.

But we can show the make-up of the marginal cost of extraction in Figure 2.1. Suppose we are in period t. The lowest marginal cost curve, MC_t, is the per-unit effect of extraction in period t on cost in that period; this can be expressed as $\Delta C_t / \Delta q_t$, where Δ is "the change in." As is usually the case, marginal cost increases with output over the relevant range. The second curve is MC_t plus the discounted sum of the per-unit effects of extraction in all future periods until the end of the horizon, H. This part of marginal cost is expressed as $\Delta C_i / \Delta q_t$, where the i-th period is some period beyond t up to H. The highest marginal cost curve includes the opportunity cost or user cost discussed above.

FIGURE 2.1

The Impact of the Future Costs of Present Output on Present Production

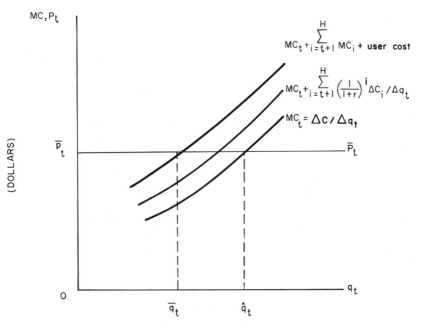

Now let us discuss the effect of these additions to marginal cost on the firm's rate of output. Suppose the firm has no effect on the price of the mineral it sells. Also in Figure 2.1 assume that the price in period t is $\bar{p}_t$. The firm would increase its rate of extraction as long as the price it receives for the mineral exceeds the marginal cost of extracting the mineral. It would not increase its rate of extraction if the marginal cost exceeds the price. Thus, in equilibrium price equals marginal cost. In Figure 2.1 if only current marginal cost, MC_t, is considered, the firm would extract $\hat{q}_t$ in period t. But when all effects are considered in the marginal cost curve, output in the t-th period declines to $\bar{q}_t$. In other words the effect of current extraction on future costs and future profits decreases current output below what it would be in the absence of these future effects, as would be expected. Because of these future effects the firm is forced to conserve resources for the future.

One must note, however, that increases in the relevant rate of interest tend to increase present rates of extraction relative to future. That is, increases in the interest rate cause present profits, which can be invested at the higher rate, to increase in value relative to future profits, which are more heavily discounted. Therefore, the rate of exploitation of known deposits tends to be increased. On the other hand, the exploration, development, and investment are affected by the rate of interest. For this reason there may be some offsetting forces to the increased rates of extraction of known deposits. The direction of change depends upon the strengths of all forces. We will discuss exploration, development, and investment at greater length below.

Before turning to more complex models, we will set forth this simple model mathematically. The nonmathematical reader may omit this exposition and skip to the next subsection. In mathematical terms the firm maximizes present value or net worth, written as

$$PV = \sum_{t=0}^{H} \left(\frac{1}{1+r}\right)^t [P_t q_t - C_t(q_t, q_{t-1}, q_{t-2}, \ldots q_0, Z) - F_t] \quad (2.5)$$

subject to the constraint

$$Q = \sum_{t=0}^{H} q_t,$$

where P_t is the expected price in the t-th period
           r is the relevant rate of discount
           C_t is the cost function as set forth above
           F_t is fixed cost in the t-th period
           Q is the total amount of the mineral deposit

The constraint merely says that the firm can only extract Q and no more. It will in this model extract no less than Q.

The prices in each period are assumed to be given to the firm. Therefore, the choice variables are the amounts to be extracted in each period. Maximization of present value requires that for each period t

$$\left(\frac{1}{1+r}\right)^t \left(P_t - \frac{\partial C_t}{\partial q_t}\right) - \sum_{i=t+1}^{H} \left(\frac{1}{1+r}\right)^i \frac{\partial C_i}{\partial q_t} - \lambda = 0, i = 0 \ldots H \quad (2.6)$$

where λ is the Lagrangian multiplier or the shadow price of an additional unit of the deposit. Thus, maximization of present value requires that in each period the discounted expected price equals the discounted total marginal cost.

First, there is the effect of current output upon current costs, the typical effect, expressed as $\partial C_t / \partial q_t$. Second, there is the discounted effect of current extraction upon the cost of extraction in subsequent periods; these are expressed as

$$\sum_{i=t+1}^{H} \left(\frac{1}{1+r}\right)^i (\partial C_i / \partial q_i).$$

These costs can vary over the horizon. The cost effects must be discounted at the relevant rate. Finally, there is the "user cost" of current extraction, expressed as λ. This cost shows the opportunity cost of current extraction on future profit. A unit of output extracted now cannot be extracted in the future. Thus, both the second and third effects tend to increase marginal cost in the current period above the level that would result if marginal cost depends only on current output. Therefore, output is shifted to the future. This exposition merely sets forth formally both what was discussed above and the results shown in Figure 2.1.

Some Additions to the Model

The model set forth above is the simplest model of the mineral-extracting firm. Clearly, no firm would consider its resource deposit as totally fixed in supply. As we see time and again, a reserve at one price is not a reserve at a lower price. Future price and future cost determine the total amount of the deposit that will be extracted. We see potentially profitable reserves becoming uneconomical if the cost of extraction is expected to rise or if the price of the mineral is expected to fall. This complication may be incorporated by changing the constraint from $\Sigma q_t = Q$ to

$$P_{H+1} q_{H+1} < C_{H+1} + F_{H+1} \qquad (2.7)$$

This constraint eliminates the assumption of a fixed resource deposit and requires that when revenue in any period becomes less than cost in that period, the firm will shut down. It would continue extracting if price rises or cost falls in the $H+1$ period or beyond. Using this constraint, the opportunity or user cost—the third element in marginal cost—becomes $\lambda(1/1+r)^{H+1}(\partial C_{H+1}/\partial q_t)$. That is, the user cost is the present value of the increase in cost after the final period of production due to extraction in the current period. In this case the firm has an additional decision to make: determining the end of the time horizon, or when to abandon the deposit. But this complication does not particularly change the conclusions of the basic model.

Many other attempts to make the theory more realistic would lead to additional complications. While we have thus far assumed that only one deposit is being exploited, it is clear that firms extract from several deposits simultaneously. But this extension does not significantly affect the analysis. One could overcome the problem by simply assuming that the firm maximizes the present value of each deposit it is exploiting and therefore maximizes the sum of the present values as set forth above. Or, more easily, one could assume that the above present value function applies to all of the deposits in question. That is, the relevant q_t stands for the total amount extracted from all deposits. We should also note that firms generally extract different types of minerals, often from the same orebody. This occurrence need not give particular problems. A more complicated extension is the appearance of different qualities of the same type of minerals. Again, one may simply interpret output as some composite amount of extraction consisting of different qualities of ore and of different minerals. The cost function would require adaptation in order to take these different forms into consideration. But there is little or no change in the basic model itself. For our purposes it is generally not necessary to make these complicating modifications.[2]

However, there are places in our analysis where we will extend the model to cover different grades of ore in the same deposit. For example, as we show below, one method that firms can use to adapt to changes in taxation or in environmental regulations is to change the grades of ore they extract. For example, some types of taxation can lead to hygrading, i.e., raising the quality of the lowest grade of ore extracted.[3] For the most part, however, unless we are specifically interested in this aspect of the problem we shall assume that minerals in a deposit are of uniform quality. This assumption does not generally change the results of our deductions. In fact, for the most part we will be interested in the change in investment and in total output of all ore resulting from changes in institutional factors.

Finally, we could consider the type of firm that both extracts minerals and operates its own smelter. Given competition, it would be profitable for an extracting firm to operate its own smelter only if positive externalities exist or if smelting is profitable in and of itself. But in the long run under competition in each sector, both extraction and smelting would earn only the prevailing rate of return on investment. In this case, in the absence of cost advantages, a firm would build its own smelter at an excavation site only if it could obtain a substantial cost reduction by doing so. Otherwise, the firm would merely ship its ore to a separate smelter. Certainly, the orebody would have to be large enough to make the combination worthwhile. The economics of vertical integration would be quite likely to come in the area of transportation, reduced transaction costs (e.g., scheduling), and possibly managerial economies. The basic model set forth above can take account of the possible reduction in cost from carrying out both operations at the site.

We could consider the implications of firms monopolizing both stages of production, or we could examine the welfare effects or social benefits of measures encouraging extracting firms to carry out their own smelting operations in the same economic region. Neither discussion, however, would add to the basic analysis of this work. Let us, therefore, turn to a subject that will be of extreme importance throughout much of the remainder of this book: investment, exploration, and the effect of uncertainty.

INVESTMENT, EXPLORATION, AND UNCERTAINTY

Until this point we have assumed extraction from a known orebody under conditions of relative certainty. But this gives us only one aspect of the firm's activities. Mineral firms not only extract from given, known deposits; they also explore and develop new sites for future extraction. They invest in new capital equipment, both in deposits already being worked and at new deposits. They carry out research and development in order to discover new technologies and new techniques of production.* All these activities are undertaken under a certain amount of risk or uncertainty.

Let us begin our analysis by lumping all such types of activities into one framework and call it *investment*.[4] We first analyze the investment decisions of a firm without considering the risk or the uncertainty involved. Then we examine the problems caused by introducing risk and uncertainty.

Investment involves the use of scarce resources and hence can be carried out only at a cost. The total cost of investment undertaken in a particular period depends upon the type and the extent of investment carried out. Investment of any type is undertaken only if it is expected to yield a return in excess of cost.

Assume first that the total expenditure on investment by a firm during a period depends upon the total amount of investment in that period. The marginal cost of investment is the additional cost resulting from an additional unit of investment. Either of two assumptions may be made about this marginal cost. It can be assumed that the marginal cost of investment is constant over the relevant range of investment in a period. This is the case if units of investment are measured as dollars spent on exploration. By definition, the marginal cost of an additional dollar's worth of investment is one dollar. Alternatively, units of investment can be measured in other terms— such as man hours spent and so on—and the marginal cost of investment can be assumed to increase with increased investment. As is apparent below, the latter assumption unduely complicates the problem. Thus it is assumed that investment is measured in dollars and the marginal cost is constant.

---

*The question of the assimilation of new technology is addressed in Chapter 5.

Next assume that potential investment activities can be ranked according to expected returns from the highest returns to the lowest. These returns are the expected discounted streams of returns from the investments. Call the marginal return to investment the expected increase in present value for an additional unit of investment. From our assumption about ranking, the marginal returns decline with increased investment activity. Clearly, the firm undertakes investment as long as the marginal expected return exceeds the marginal cost.

This conclusion is illustrated in Figure 2.2 in which it is assumed for simplicity that investment activity is a continuous function. The marginal cost of investment, assumed constant, is OM. If investment is measured in dollar expenditure, OM equals one dollar. MR is the expected marginal return from additional units of investment; the negative slope of MR reflects the assumption that the activities can be ranked from the highest to the lowest expected return. The firm undertakes $O\overline{E}$ investment activity during the period. It does not invest less than $O\overline{E}_t$ because the expected return from additional investment exceeds its expected cost. For investment beyond $O\overline{E}_t$ the added cost exceeds the added revenue.

Changes in certain exogenous variables may alter the amount of investment activity undertaken. For example, an increase in the rate of interest would decrease the present value of the expected stream of income and therefore shift the marginal revenue curve downward. Thus, an increase in the interest rate would decrease investment. A decrease in the rate of interest would have the opposite effect. Alternatively, technological change would make resources more productive and therefore shift the MR curve outward. Other factors could affect the MR curve also.

Another method of analyzing the optimal level of investment for a firm is to use rates of return rather than streams of return. Using this method of approach in Figure 2.2, the horizontal axis would still be the level of investment in a particular period. But the vertical axis would show the *marginal rate of return* on investment at each level of investment. The curve showing the marginal rate of return would still slope downward because of the assumption that investments are made in order of their rates of return from high to low.

A firm in any period would undertake investments so long as the return from each investment exceeds the relevent rate of interest. Suppose for example that six different investments are available to a firm. These yield in order returns of 12, 10, 8, 6, 5, and 4 percent. If the rate of interest is 7 percent, the firm would undertake the first three investments only. If it could borrow at 7 percent and make more than a 7-percent rate of return, the investment is profitable. Put another way, the interest rate (here 7 percent) is the opportunity cost of making an investment, because that is the return the firm could earn if it did not carry out the investment.

FIGURE 2.2

Optimal Level of Investment

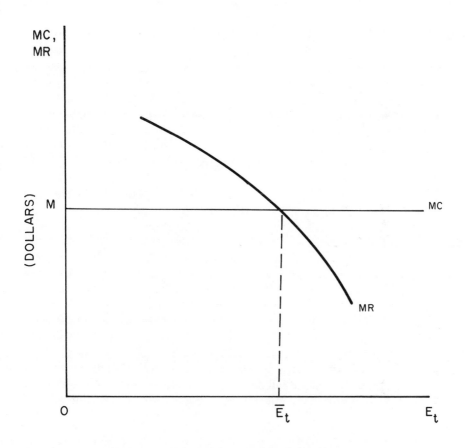

(LEVEL OF INVESTMENT MEASURED
IN DOLLARS IN PERIOD t)

For simplicity we frequently assume that the investment opportunities are continuous. Thus, if in Figure 2.2 rate of return is plotted on the vertical axis, then MR could represent the marginal rate of return at each investment level. Let OM be the rate of interest. Clearly, the firm would invest so long as the marginal return exceeds the interest rate and therefore would spend $O\overline{E}_t$ on investment. As before, it is apparent that higher interest rates decrease investment and lower interest rates increase it.

We can now examine briefly the effect of introducing uncertainty into the model. Assume that the expected return from an additional investment activity is

$$ER = R + \varepsilon \qquad\qquad (2.8)$$

where R is the expected value of the return and ε is a random variable. Assume that the expected value of ε is zero, and that ε has a nonzero variance. That is, the investor recognizes that the return can be greater or less than the expected value. If the firm or potential investor is risk neutral, the model described above, which ignores uncertainty, still applies. That is, exploration would be undertaken so long as the expected value of the marginal return equals the marginal cost.

But if the investor is, as is frequently assumed, risk averse, the conclusions are altered somewhat. An investor who is risk averse places a lower value on R, the greater the variance of the random variable ε, even though the expected value of ε is zero. That is, the greater the uncertainty about the expected stream of returns or the rate of return, the lower the value placed by the risk-averse investor on the expected stream of income.

Therefore, uncertainty from the viewpoint of the firm causes the MR curve to shift downward. The strength of the shift depends upon the discount factor that the firm places on risk. If it is assumed, as seems reasonable, that ε, the random variable, is distributed normally, then anything that increases the variance of ε, increases the downward shift of the MR curve and therefore decreases the amount of investment undertaken. Therefore, if investors are to some extent risk averse, activities that increase uncertainty decrease the amount of investment undertaken by firms and hence by the industry. This is an important point that we will discuss to a greater extent in our analysis of taxation and controls in the mineral-extracting industry. Uncertainty plays a large role in that analysis.

INDUSTRY SUPPLY AND DEMAND

The world price of a mineral, and in effect the price in a specific country or portion of the country, depends on supply and demand. We are justified in speaking of a world price of a mineral if resources are permitted to flow among countries. Certainly, the price of a mineral can vary between two mineral exchanges in two countries at a specific time. But, it cannot vary by much more than the cost of shipping the mineral for very long, since people would buy the mineral in the country with the lower price and ship it to the country with the higher price, thereby driving the price down in the one country and up in the other. Of course, in different countries there can be artificial restrictions, such as import quotas, tariffs, agreements between governments, and so forth, which could eliminate arbitrage. However, in general we can speak of a world price of the mineral, which is, in large part, determined by supply and demand.

We might note that the remainder of this section and the next section consists of rather basic economic theories of supply, demand, industry equilibrium, and rents. The discussion is primarily for the benefit of readers who are not economists. They may be omitted without loss of continuity by professional economists who are generally quite familiar with the analysis.

Demand

The demand for most minerals is a derived demand, since minerals are generally used as inputs in the production of other goods rather than purchased by the ultimate consumer. As with all demands, the quantity demanded of a mineral varies inversely with its price. This inverse relation is due to the substitution effect and the output effect.

Consider first the substitution effect. As the price of the input rises, firms will attempt where possible to substitute other inputs. While there are some productive processes in which some inputs have no substitutes, in the majority of cases at least some substitution is possible. To use an extreme example, if the price of steel becomes too high, some automobile manufacturers may substitute fiberglass for some parts that were previously made of metal. There are many other not-so-extreme examples. Exactly the reverse occurs following a price decrease. The input is now relatively cheaper and firms tend to substitute toward its usage. We will examine extensively in Chapter 4 the substitutability of inputs in the minerals industry.

The output effect, resulting from an increase in the price of an input, causes an increase in the production costs of those firms using the inputs. This increase in costs results in an increase in the price of the good produced with the use of the relevant input. The price increase causes consumption of that good to decline, thereby resulting in a reduction in the use of the input in question. Again, the opposite occurs following a price decline; the usage of the input increases as the costs of the produced product fall, and its consumption rises. In both cases the output effect reinforces the substitution effect, and thus price is inversely related to output for an industry.

The degree of the inverse relation, called the *elasticity of demand*, measures the relative responsiveness of quantity demanded to changes in price. The more responsive the quantity demanded to price, the more elastic (less inelastic) the demand. The elasticity of derived demand depends on several factors. The greater the elasticity of demand for the product produced, the higher the elasticity of the derived demand for the input. This result follows because a high-demand elasticity means that output of the product declines to a greater extent when the price of an input causes costs to increase and the price of the produced good to increase. Thus, the usage of the input decreases greatly, and the output effect is substantial. The analysis is easily extended to a price decrease.

It is rather obvious that the strength of the substitution effect varies directly with the ease of substituting other inputs in the production process. If other factors of production are readily substitutable, the usage of an input declines greatly after an increase in its price. On the other hand, inputs for which it is technologically quite difficult to substitute would not experience a particularly large decrease in usage.

The elasticity of supply of substitute inputs measures the degree to which the price of these inputs increases as the industry increases its demand for these inputs. For example, suppose that two metals can be substituted in the production of a particular good. Let the price of metal A rise. Users are induced to substitute metal B for A to some extent. If increased usage of B causes the price of B to increase substantially, substitution is carried out to a lesser extent than if the increased usage causes little or no increase in the price of B. Thus, the more elastic the supply of B, the more elastic the derived demand for A.

Finally, there is the effect of the time period of adjustment upon derived demand elasticity. While discussion has thus far abstracted from the effect of time, the time period of adjustment can be quite significant. For example, in most production processes it takes a substantial period of time to carry out a major change of the method of production after a change in the relative prices of certain inputs. Firms are set up to produce using specific techniques. These techniques cannot generally be altered overnight. But in the long run they can be altered when the price of one factor of production or ingredient input changes relative to others. There is a preponderance of empirical evidence that, given a time period of adjustment, firms do substitute away from an input that becomes relatively more expensive.*

Similarly, the output effect takes a certain amount of time to take place. Consumers do adjust their purchases in response to cost-induced relative price changes, but the adjustment is frequently more significant after a certain period of time has elapsed. In fact, producers who use the input in production do not immediately change the price of their product after a change in the price of the input. While the market works to allocate resources, it does not always work instantaneously. For these reasons it is expected that the derived demand for an input becomes more elastic the longer the time period of adjustment.

---

*Some empirical evidence is presented in Chapter 4.

Supply

Just as in the case of mineral demand, we can speak of world supply of a mineral, the supply of a particular nation, or the supply of a specific part of the nation. At any time the amount supplied by an industry is made up of the amounts supplied by member firms. In practically all cases the amount of minerals supplied varies directly with the price of the mineral. Thus in general, if society wishes more of a mineral, price must increase.

If price rises, quantity supplied increases for several reasons. First, existing firms will try to produce more from deposits already being worked. This type of expansion may involve increasing input prices. If such is the case, the increase in output would fall short of the increase that would have been forthcoming from all of the firms if the input prices and hence costs had not been bid up by the industry.

The quantity supplied can also increase by increasing the number of firms or the number of orebodies worked. For example, if the price of a mineral increases, and the increase is believed to be relatively permanent, new deposits, not believed profitable at the lower mineral prices, may now become profitable. The amounts extracted from these new deposits are subsequently added to quantity supplied. Finally, the higher price can occasion increased exploration and development, thereby leading to new discoveries and possibly to increased output. All these forces, resulting from price increase, work similarly but in the opposite direction in the case of a price decrease.

Thus, there are basically three factors that affect a mineral industry's elasticity of supply—that is, the relative responsiveness of output to price changes. First, there is the state of the technology and the way in which industry output affects the costs of input. These forces affect the cost structures of the extracting firms and therefore the total amount supplied at each price.

The second factor affecting the elasticity of supply is the ease or cost of bringing previously unworked orebodies or deposits into production. The more costly it is for firms to begin producing from these additional deposits, the greater the increase in mineral price necessary to make additional, previously unprofitable deposits, profitable. If the cost of bringing these previously unworked deposits is high, then the price of the mineral must increase substantially before these new deposits can become profitable, and the supply of the mineral will be relatively inelastic. If, on the other hand, a slight price increase makes many additional deposits profitable, the supply will be quite elastic.

The third factor affecting the elasticity of supply is the time period of adjustment. Other things being equal, the longer the time period of adjustment, the more elastic the supply. For example, suppose the price of a mineral increases and the increase is expected to be relatively permanent. At first there can be relatively little increase in output in response to the increase in price. It simply takes time to increase output. But given more time to adjust, the new higher price induces firms to work the orebodies already in production more extensively. As noted above, the higher price can make previously unprofitable deposits profitable and cause an additional increase in output.

If the price increase causes additional exploration in new areas and if the exploration is successful in finding new deposits, output can expand even further. Because of the uncertainty involved in exploration, its effect on supply cannot be specified rigorously. But, the general description gives some indication of the effects of time. Supply becomes more elastic the longer the length of time during which the industry is allowed to adjust. We might note that the adjustment is similar but in an opposite direction for a price decrease. The length of the time period of adjustment of course varies from industry to industry.

Industry Equilibrium

The world price of a mineral is determined by the intersection of supply and demand, if the industry is not characterized by monopoly. There have been cases in which the output of one firm greatly influenced the world price of a particular mineral, but in general most mineral industries behave somewhat similarly to the competitive model.[5] When price is below equilibrium, quantity demanded exceeds quantity supplied, and there is a *tendency* for price to be bid up by demanders. When price is above equilibrium quantity supplied exceeds quantity demanded, and there is a *tendency* for price to be reduced by suppliers.

For equilibrium price and quantity to change, something must cause either supply to change (possibly a new technology or a new ore discovery) or demand to change (possibly from a technological change in an industry using the mineral or a change in consumers' tastes patterns). But the effect of such changes is not felt instantaneously. In fact, the effect can be spread over a considerable period of time.

For example, suppose a technological change in manufacturing increases the productivity of, and therefore the demand for, some mineral. If demand increases, there would probably be little or no immediate effect on output. But demanders, bidding for the available output, will drive up the price of the mineral, causing increased profits for the mineral firms. The higher price and the increased profits lead to expansion of the rate of extraction from orebodies currently being worked. Output would increase, and price would fall somewhat.

In the long run new deposits become profitable and can be brought into production, leading to an increase in output and a further decrease in price. Finally, if the increased profits cause increased exploration and development, output increases more and price decreases. Price will probably end up somewhat higher than the original price, but this is caused by the increased demand. If in fact the profit-induced exploration and development is extremely successful, output may increase so much that price need not rise at all. In any case, the market over time adjusts to the increased demand for the mineral by enticing greater output from the mineral firms.

There are, however, conceivable circumstances under which the increased demand does not lead to an increase in output. It is possible that output cannot physically be expanded, but this circumstance seems remote: History records no such example. More likely, it may be the case that the price rise is insufficient to make other deposits profitable. It is not physically impossible to increase output; it is simply not profitable. A higher price would perhaps lead to a greater output.

Another circumstance in which an increase in demand may not bring about an increase in output involves governmental intervention. Governments may at some stage become worried about "excess profits" being made by firms in a particular country or in a particular region of a country. In this case government can do one of two things, either of which will prevent the increase in quantity supplied because of an increase in demand. It can impose a ceiling price, preventing price from rising above a specific level: thereby there will be no incentive for the industry to produce more output. Alternatively, government can tax away "excess profit." In this case the higher price does not lead to more profits, and firms have no incentive to increase the rate of extraction. Firms react to the inducement of after-tax rates of return. If they recognize that additional before-tax profits are to be taxed away, it is unlikely that they will attempt to increase the rate of extraction. This effect will become more clear in our exposition of the effects taxation.

THE FIRM AND INDUSTRY IN THE LONG RUN

Until now we have simply assumed that equilibrium requires that each firm choose its profit-maximizing rate of output and that, for the industry, quantity supplied equals quantity demanded. In this section we will examine some of the implications of long-run equilibrium for a mineral-extracting industry in a specific economy. We begin with an analysis of the way in which profits are competed away in the long run; then we discuss some of the implications of rents in mineral industries.

Long-Run Equilibrium

In our earlier discussion of investment we deduced that firms in an economy will undertake various types of investments until the rate of return from the marginal investment equals the relevant rate of interest or the prevailing rate of return in the economy as a whole. Thus, if there are profits in one part of the mineral-extraction sector of an economy, resources will be attracted into that sector until the return to the marginal investment is driven to the prevailing rate of return.

In the process the returns (profits or rents) to the previously profitable firms may or may not be competed away, depending upon whether the firms are price takers in the world market for the mineral. If the mineral-extracting sector of the economy has no effect upon the world price of the mineral, expansion of this sector will not drive down the price of the mineral received by previously profitable firms. But there is still the tendency for the profits from marginal investments to be competed away. If the extraction rate of an economy does have an effect on the world price of the mineral, the price received for the mineral will fall, and the profits of previously profitable firms will be driven toward the prevailing rate for the economy as a whole. This is not to say that all returns that are above normal will be competed away, but there is a tendency for the average returns in various industries to be equalized. When one industry shows consistently above-normal returns, investment will be attracted into that industry, and returns will be driven to normal. Alternatively, when returns are consistently below the normal rate of return, investment will be withdrawn from the industry, and returns will increase to the normal rate. Thus, in the long run there will be a tendency for returns in industries in an economy to be equalized.

Rents

We will conclude this section with a brief discussion of the concept of rent. To begin, assume that the long-run industry supply of some mineral is given; demand is given, too, and an equilibrium price and quantity are determined. The supply curve is derived under the assumption that it shows the *minimum* price necessary to elicit each level of output. Any price to a producer above the minimum necessary to bring forth this level of output leads to rent, which is a return above the alternative cost of capital.

Now let us determine the recipient of the rent. Suppose that there are a large number of mineral-extracting firms. Also assume that someone owns a property that everyone knows will yield a return above the going market price. If the exact prices and costs over time are known with certainty by all parties, the mineral firms through competitive bidding will drive up the royalty and bonus payment until all rents are competed away for the firm that receives the lease. The landowner will obtain all the rents. Thus, if the initial supply prevails under certainty, all intramarginal landowners receive rent; the amount depends upon the difference in revenue and costs, including the opportunity cost of capital. The owners of marginal lands will receive a royalty only sufficient to cover the alternative cost of the land in the next best opportunity. In fact, strictly speaking, only the payment to owners of intramarginal land above the royalty necessary to cover its alternative cost can be spoken of a true rent. Thus, with no uncertainty, in the long-run with competition the firm receives no true rent above the going rate of return on capital.

It goes without saying that complete certainty does not prevail in mineral-extracting industries. In fact, it does not prevail in any industries, but competitive forces *tend* to eliminate rents in the long run. For all potential properties not being exploited, mineral firms and landowners have some notion of the potential worth of the properties. In the absence of certainty all the rent may not necessarily go to the landowners. Clearly, firms bid for the property. But the expected yield may differ considerably from the realized yield. In addition, the division of the realized yield depends upon the division of land payment into royalty and lease bonus; such division depends considerably upon the risk aversion of landowners relative to firms. The more risk averse the landowners, the more likely for rents to be largely in the form of lease bonuses. In this case, the reward for yields above average go to firms; alternatively, the costs of yields below average are borne by the firm also. Even if government is the landowner, the above analysis is little altered, but the government receives the royalty in the form of taxation. Who bears the risk depends upon the risk averseness of government relative to that of firms.

Just as an aside it is worthwhile to examine the relative distribution between bonuses and royalties when private individuals are the prominent landowners and when government is. It is a generally accepted economic hypothesis that the larger and more diversified the economic agent, the less risk averse that agent. In addition, since the larger party can generally diversify more economically, it is better from a social sense that the larger party in a business venture bears the risk.

Thus it would appear that in transactions between a firm and a private individual—under competition—the firm would bear the majority of the risk since it is generally the larger party. In transactions between government and a firm, the government would bear the majority of the risk. One might, therefore, deduce that the proportion of royalties to lease bonuses is higher when government is the landowner. (Of course, this assumes that government is a profit maximizer in such transactions; such may or may not be the case.)

In any case, regardless of the type of payment made for leases or who receives the rent, under competition there is, in the long run, a tendency for the average rate of return in a mineral industry to be approximately equal to the going rate of return in the economy as a whole. This is not to say that some firms during this period may not be earning above-normal rates of return. But some are also earning below-normal returns. So long as firms are earning above normal returns in an industry there is entry into new, perhaps marginal areas or deposits. This increases supply, which of course drives down the price to all firms. Again, some firms with favorable cost conditions and possibly favorable lease situations continue to earn rents. But there are some not-so-favorably situated firms that earn no rents and may not even cover their costs. If entry continues too far, or if market conditions become unfavorable and returns generally begin falling, the marginal producers exit, decreasing supply until returns rise.

This then is the basic theory of the mineral-extracting industry in the long run. Our theory assumes that the industry is reasonably competitive and that the rate of return on capital will approximately equal the rate of return on capital in the economy as a whole. Most of the empirical material in the chapters that follow is based upon or makes use of these important assumptions. Therefore, before proceeding it will be useful to show that from the evidence we have the average rate of return in the mineral industry, and have in fact not differed significantly from the average rate in manufacturing in the economy as a whole. This is shown in several ways in Chapter 3. While the absence of "excess profits" in minerals is not necessarily evidence of competition in minerals, it does give us a reason to use the competitive model because of the absence of evidence to the contrary—that is, the absence of an above-normal average rate of return. Thus, we will continue to use the model described here.

SUMMARY

This chapter has developed the basic model to be used to analyze the behavior of the mineral-extracting firm and industry. In this model it is assumed that the firm maximizes its present value or net worth. The model recognizes explicitly the dynamic condition inherent in mining that extraction in the current period increases the cost of extraction in future periods. That is, the more ore extracted now, the greater the cost of extracting a specific amount of ore in a future period. The firm is thereby faced with two related decisions. The first, the rate at which to extract ore, is determined by equating discounted expected price with discounted marginal cost, which includes not only the effect of current extraction on current cost but also the effect of current extraction on future extraction costs and profits. The second decision, the length of the period over which to extract the ore (or the total amount of the reserve to extract) is dependent on expected future prices of the ore and the effect of current extraction on the future costs of extraction.

The level of investment by the mineral-extracting firm—whether in the form of capital investment, exploration, or research and development expenditures—is determined by equating the marginal cost of and the expected marginal return from the investment. In this activity, if firms are risk averse and are faced with uncertainty, they will invest less than they would under certainty. However, if the firm is allowed to share risk with other firms or with the government, the level of investment will be higher than in the absence of this risk pooling.

In our discussion of the mineral extracting industry, demand, supply, and market equilibrium were analyzed. It was noted that, since the demand for the mineral is a derived demand, its price elasticity (responsiveness) depends on the elasticity of demand for the final product that uses the mineral as an input, the technological ease with which other inputs may be substituted for the mineral in the final production process, the supply elasticity of these substitute inputs, and the time period available for adjustment. The price elasticity of the supply of the mineral is determined by the state of technology, the responsiveness of input prices to changes in wage level by the mineral-extracting industry, the availability of additional deposits, and the length of the time period under consideration. It was shown that in the mineral-extracting industry the adjustment to the final market equilibrium may take a considerable amount of time, because adjustments in supply would come about first in an expansion of the rate of extraction from existing deposits, then from production from additional deposits, and finally through exploration and development. Nonetheless, in the final equilibrium position, our assumption that the mineral-extracting industry is competitive would mean that the rate of return in mineral extraction would approximate the rate in the rest of the economy, and rents should be competed away.

In this chapter we have mentioned the relevance to our theoretical model of our assumption of competitiveness, the form of the underlying production function, and the state of technology. These issues will be addressed explicitly in the following three chapters. In Chapter 3, data from both Canada and the United States will be considered to determine whether the existing evidence is consistent with our assumption that the mineral-extracting industry is competitive. Chapter 4 will present some empirical evidence concerning some major characteristics of the production relation existing in mineral extraction. In Chapter 5 the basic model developed here will be extended to consider the conditions under which a mineral-extracting firm will employ some unconventional extraction technology, the effect of this technology on the supply of minerals and on the market equilibrium, and the economic feasibility criteria for employing new technology.

NOTES

1. This point was noted by Anthony T. Scott in his paper, "The Theory of the Mine under Conditions of Certainty," in Mason Gaffney (ed.), *Extractive Resources and Taxation* (Madison: University of Wisconsin Press, 1967) 25–62. A discussion of this point is found in the appendix to this chapter.

2. As pointed out in Frederick M. Peterson and Anthony C. Fisher, "The Exploitation of Extractive Resources: A Survey," *The Economic Journal* (December 1977): 681–721, this problem of multiple deposits of different grades of ore has been addressed explicitly by Martin Weitzman ("The Optimal Development of Resource Pools," *Journal of Economic Theory* [June 1976]: 351–64), Robert Solow and Frederick Wan ("Extraction Costs in the Theory of Exhaustible Resources," *The Bell Journal of Economics* [Autumn 1976]: 359–70), and James Sweeney ("Economics of Depletable Resources: Market Forces and Inter-temporal Bias," *Review of Economic Studies* [February 1977]: 125–42).

3. This follows from Carlisle's observation that the amount of the total deposit extracted and the cut-off grade is an endogenous decision made by the firm on the basis of profitability: (Donald Carlisle, "The Economics of a Fund Resource with Particular Reference to Mining," *American Economic Review* (September 1954): 595–616). Carlisle's approach is described in the appendix to this chapter.

4. In this analysis we are primarily concerned with the investment activities in general rather than in any particular type of investment. For a review of the literature concerned specifically with exploration and with the effects of uncertainty, the reader is referred to the survey article by Peterson and Fisher, *op. cit.*

5. This point is addressed in Chapter 3. For a consideration of the effects of monopoly, the reader is again referred to Peterson and Fisher, *op. cit.*

APPENDIX

HISTORICAL DEVELOPMENT OF
THE THEORY OF
THE MINERAL-EXTRACTING FIRM AND INDUSTRY

Long before the recent surge of interest in the economics of mineral extraction, or the economics of exhaustible resources, this area of study had a rather long and respectable history. Many extremely good papers concerning mineral economics appeared long before the current running-out-of-natural-resources scare. Of course, within the past few years a plethora of articles and books of widely varying interest and significance have entered into the literature. Therefore, the number of works in natural resource economics in general or in specific areas of this field is extremely large.

Because of the great number of articles and books in the mineral-extraction field, a review of the literature must take one of two forms. The review can either be encyclopedic, summarizing briefly practically everything that has been written, it can be selective and set forth a thorough exposition of several "key" works, on which most other papers are based. We take the latter approach.

Furthermore, in this discussion we will limit our attention to those works that form the basis for the general theory of the mineral-extracting firm and industry as it is currently received. A large number of works deal with extensions of the basic model in order to consider such issues as the effect of monopoly power in the mineral-extracting industry, recycling, the optimal rate of consumption of a nonrenewable resource, and the question of intergenerational welfare. However, since these topics are tangential to our analysis we will not discuss them. The reader interested in these extensions is referred to the survey article by Frederick M. Peterson and Anthony C. Fisher.[1] While this appendix focuses primarily upon the economics literature, interested noneconomists should be able to follow most of the material rather easily.

THE THEORY OF THE MINERAL-EXTRACTING FIRM

Most of the theory of the mineral extracting firm is concerned primarily with the optimal rate at which some given resource deposit or resource field should be extracted and secondarily with the optimal total amount that should be extracted. There is a good deal of agreement in the basic writings. The major differences are in the types of assumptions made.

Probably the first paper in the area, and still a classic, was one by Lewis C. Gray, appearing in 1914.[2] Although incorrect on certain points, this paper was a forerunner of many later writings on the theory of exhaustible resources. While Gray was basically interested in the taxation of rents from mineral extraction, he developed a simple theory of the mineral-extracting firm as an aid to his analysis. Gray began analysis by noting that where land, or its properties, are inexhaustible, land is like labor because its product is lost when not used. Alternatively, when the properties are exhaustible, as in the case of a coal deposit, use diminishes the benefits.

Gray noted that the owner of a deposit might well maximize his profits by postponing extraction into the future, perhaps because of an expected increase in the price of the resource or an expected decline in the prices of the factors of production used in extraction. Another, more fundamental consideration, according to Gray, is that postponement of extraction is often caused by diminishing productivity of factors of production. In the typical theory of the firm a landowner maximizes his return (rent) by applying labor and capital to a given area until the additional product of the last unit of labor and capital just equals the additional cost of that unit. In other words, if a unit of an input costs $10 per unit, the owner would add units so long as the marginal product of each unit is worth more than $10; he would never add a unit that added less than $10 to total revenue. If the benefits or products of the land are exhaustible, the landowner would use less labor and capital per year than in the above case, because the additional cost of present extraction would be the sacrificed future profit.

Ignoring the effect of the interest rate and all other considerations, Gray noted that the owner of an exhaustible resource would use labor and capital only until the point of maximum average return for a unit of expense is reached. At this level of output the average net return per unit of the resource extracted is maximized. Let us emphasize that the minimization of the average expense per unit extracted maximizes returns only if the interest rate is ignored. In other words, when a dollar received in one year has the same present value as a dollar received in any other year and all prices are expected to remain constant, maximizing the return per unit maximizes total returns in the case of an exhaustible resource. If any more is mined per year, each unit will yield a smaller return than if extraction were postponed. Of course, the above analysis is based upon diminishing marginal productivity of labor and capital over a range, a concept that some writers prior to Gray denied in the case of mining. In fact, as Gray noted, even the great economic genius of the time, Alfred Marshall, did not feel that this principle held for mineral extraction. Gray himself felt that in most cases diminishing marginal productivity held, even for surface mining.

Next, Gray examined the effect of the rate of interest on the optimal rate of extraction. A positive rate of interest offsets to some extent the above-noted influence toward a reduced rate of extraction. Obviously, a dollar in the current period is worth more now than a dollar next year, because a dollar invested at the market rate of interest in year one returns more than a dollar in year two. Gray presented a numerical example showing how the optimal rate of extraction is derived. He also discussed a few modifications of the theory, such as the case in which the removal of some resource—say, coal—in one year completely changes the condition of removal in later years.

While Gray seems to have been primarily interested in the effect of taxation of rent, he did set forth a basic framework of analysis that has been followed throughout the mineral literature to the present. Even though Gray's methodology seems simple, even naive (his method of exposition is primarily the use of arithmetic examples, and he ignored exploration and discovery costs), he showed that a different method of approach was necessary when the firm extracts an exhaustible resource. This was a major step forward.

The next classic paper in the theory of exhaustible resources appeared in 1931, the writer was Harold Hotelling.[3] From his introduction it appears that Hotelling wrote to answer the conservationists' demands for regulation of the exploitation of minerals, forests, and other natural resources. He noted the general feeling prevailing at the time that the world's exhaustible resources were being extracted too rapidly and sold too cheaply. Many people were demanding regulation. Hotelling attempted to derive a theory to analyse questions such as:

(1) At what rate should a mine be exploited?
(2) How would public ownership differ from private?
(3) How should the proceeds be divided between income and return to capital?

To begin his analysis Hotelling assumed free competition and profit maximization. If 'r' denotes the rate of interest, e^{-rt} is the present value of a unit of profit at time t. Since a mineowner would be indifferent about there being a price of p_0 now rather than a price of $p_0 e^{rt}$ at time t, under competition $p_t = p_0 e^{rt}$, where p_t is the price in period t. It was assumed that p_t is a price net of the cost of extraction. Hotelling assumed that the total supply of the resource is limited to the amount 'a'. Quantity demanded, q, is inversely related to p.

As an example, Hotelling assumed the following demand for q:

$$q = 5 - p,$$

where demand is considered independent of time. Note that $q = 0$ for $p \geq 5$. As q diminishes and approaches 0, p increases to 5 at the end of extraction, time period T. It was then shown that if 'a' is the total amount of the resource, the equation

$$e^{-rt} = 1 + r\left(\frac{a}{5} - T\right)$$

represents the solution for T, the end of the horizon, and this equation has a single, positive, finite solution for T, the time of complete exhaustion. Other demand functions were shown to lead to an infinite horizon. Thus, the horizon depends upon the demand for the resource.

Hotelling then turned to the question of the maximum social value of a resource under the assumption that future enjoyment is discounted at the market rate of interest. He deduced that if the resource is exploited competitively, the rate of extraction will be the rate that maximizes social value. But Hotelling then pointed to several instances in which the optimal rate of extraction would not obtain. The first case involves the familiar common property problem. For example, if many firms can drill an oil field, no firm is induced to hold back, and the field is exploited too rapidly; moreover, oil and gas are lost. Hotelling also questioned discounting future values of utility at the market rate of interest.

He then analyzed the question of the results of monopolistic exploitation of an exhaustible resource and compared monopoly solutions to socially optimal conditions. Hotelling's primary contribution seems to have been the derivation of a basic method of approach to the problem of exhaustible resources.

Another major paper in this area, written by a geologist, Donald Carlisle, appeared in 1954.[4] Carlisle pointed out that previously economic theory had concentrated upon the rate of extraction while in reality the mining firm must decide not only upon the optimum rate of extraction but also upon the optimal amount of the total deposit to extract. No firm would ever extract the entire amount of a deposit.

Carlisle first examined the decision of the optimal rate of extraction with a given total amount of the resource. In essence the conclusions are similar to those developed by Gray. Assuming a U-shaped average total unit cost curve (ATUC), it is shown that the total profit stream is maximized by extracting at the rate where ATUC attained a minimum. The highest current rate of profit is obtained by extracting at the rate at which marginal cost equals price. This solution might occur when an extremely rapid return on capital is required. The present value of the stream of profits is maximized at some rate between the two. The lower the rate of discount, the closer the actual rate to the lower rate, and vice versa.

Next, Carlisle addressed himself to the case of alternative levels of extraction for a given area, examining the determinants of cut-off grade, workable limits of ore, and completeness of extraction. Since the first two points received only a cursory examination, we will limit our discussion to the last. Carlisle said that in a given area with a fixed rate of extraction the average total unit cost first decreased and then increased as the total level of recovery increases. Costs that are independent of total tonnage cause ATUC to decline. Other forms of efficiency are possible, but diseconomies are eventually reached.

Carlisle depicted the situation as shown in Figure 2.3. The total amount of ore recovered is plotted along the horizontal axis, and per-unit cost along the vertical. Price and the rate of extraction are assumed to be constant. Again,

FIGURE 2.3

Alternative Levels of Recovery

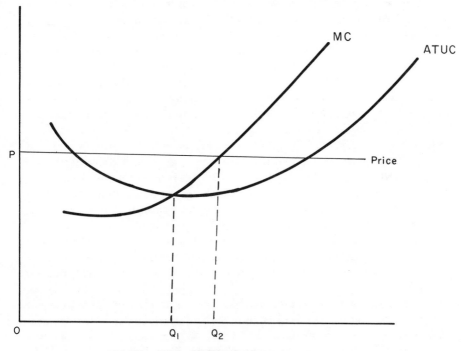

TOTAL ORE RECOVERED AT A
GIVEN RATE OF
EXTRACTION

total net receipts per unit of output are maximized by extracting the total amount at which average cost per unit of ore is minimized. Total profit per unit of time is also maximized at Q_1. But ignoring the discount rate, total profits are maximized by extracting the amount Q_2 at which marginal cost (MC) equals price. Discounted present value is maximized somewhere between; the lower the discount rate, the lower the amount extracted. Different rates of extraction under the same conditions would lead to different present values.

The two decisions are combined into one problem. The optimal rate of recovery would generally be thought to vary with the level of recovery, and the optimal level with the optimal rate. One could either compare alternative levels of recovery, each at the optimal rate, or alternative rates, each at its optimal level. In this case we would have a three-dimensional problem. The present value of the stream of profit is a function of the level and rate of recovery. Carlisle noted that optima under the double-decision criterion are not necessarily at higher rates and levels than under the assumption described above. The solution depends upon the assumptions about the way in which the optimal rate varies with the level of output.

Uncertainty, as was pointed out by Carlisle, usually complicates the problem. All technical circumstances are never known prior to excavation. He noted that some engineers have estimated that perhaps up to one-half, or even more, otherwise profitable ore might more profitably be left in the ground because of the influence of uncertainty—of depletion and discovery—upon the discount rate. In such cases it would be assumed that (1) plant capacity is small relative to total ore supply; (2) the type of mining advocated would increase the profit margins on higher grade ores; (3) a relatively high rate of discount is applied on future earnings. Of course, Carlisle stated that the highest risks are generally associated with prospecting and development, and the overall risk should not be applied to an operating mine. Obviously, there is also the risk of price changes. All the uncertainties can affect the optimal level and rate of extraction. Finally, Carlisle analyzed the possible changes when a mineral industry is monopolistic, and he examined the historical effect of price changes upon the level and the rate of recovery. In the case examined— mercury—it was found that operators tended to turn to higher-grade portions of deposits, shutting down low-grade stopes in response to a price fall. The opposite adjustment to low grades tended to lag behind price increases.

Carlisle examined many different possibilities. He thought his most useful contribution (and we agree) to be his emphasis upon the use of two-variable models in economic analysis.

About the same time that the paper by Carlisle appeared, Orris C. Herfindahl published a paper that greatly expanded upon the Gray and Hotelling articles.[5] That article dealt with three general areas of mineral economics: (1) the firm, (2) the industry, and (3) taxation. We might note in passing that Herfindahl first brought exploration costs into the theory.

The major point that concerns only the mineral firm, rather than the industry, is the analysis of the rate of extraction of a known deposit over time. Herfindahl showed that if investment in mining equipment and exploration has been made, the firm will equate the marginal net return (price minus marginal cost) to dollars of present value for each period. If this condition did not hold, the firm could increase the present value of the discounted stream of profits by shifting production between years. While this was an important point, it had been implied previously. Herfindahl's major contribution concerned the mineral industry and exploration.

Appearing in the middle 1960s was a collection of papers that carried the theory of the mineral-extracting firm several steps forward. One of the more important was by Anthony T. Scott, who was concerned with the firm's optimal rate of extraction from a given, known deposit.[6] Whereas most of the previous literature had been concerned with setting forth the final stable equilibrium path, this paper analyzed the periodic decision making of the firm to see whether the process lead to the theoretical equilibrium.

He assumed (1) certainty of reserves, (2) a constant rate of interest, (3) maximization of present values, (4) that conditions that affect cost (uniform grade) are constant over time, (5) that the firm is a price taker, and (6) that the rate of extraction can be changed only within the capacity of the present plant. It was also assumed that in each period the average and marginal cost curves are U-shaped.

Scott noted that in any current period the firm has a profit curve and a user cost curve, indicating the present value of future output sacrificed by each unit of output extracted in the period. Not only does current production use up the deposit and limit future production; current output affects future production costs. While the initial output can lower future cost as the mine is established, beyond some point current output raises future costs. The mineral-extraction firm thus faces not only a marginal cost of producing mineral output in the present but a marginal cost generated by the impact of current extraction on future production. The presence of user cost induces the mining firms to extend the period of extraction beyond the level that would exist in its absence.

If a mine contains several different grades, the problem is complicated enormously, unless there are uniform deposits of different grades separate from each other. In this case the better grade will be mined first. Other things being equal, the better the grade, the more net income that can be made. The present value of a stream will be higher, the earlier the higher profits can be invested. Alternatively, as Scott noted, the firm can gain by reducing quickly the amount of natural capital tied up in nonliquid form. When grades are mixed or are not equally accessible, the problem is not so susceptible to a generalized theory and usually lends itself only to ad hoc analysis. Scott showed that rate, grade, blend, and sequence of rate, grade, and blend must be planned by the firm on a continuing basis.

Changes in input prices and the price of the output also affect the rate of extraction. In the first case expected increases in future resource prices will decrease current output, since user costs will increase. In fact, if prices are expected to rise a great deal, user cost may rise above profit at every level of current output, and all production would be moved into the future. The effect of input price changes will be in the opposite direction. An expected increase in wages, for example, would tend to raise future costs, decrease future profits, reduce user costs, and increase current extraction. Again, a sufficient decrease in expected wages could cause all present output to be conserved. Scott extended the analysis briefly to show the effect of taxes and the choice of a particular level of investment. The analysis of these problems is given later in this book.

Appearing about the same time as the Scott paper was a paper by Richard L. Gordon.[7] Using rigorous mathematical analysis, Gordon developed a model of a mineral-extracting firm and from this model reached several fundamental conclusions. He assumed that the firm has available a certain resource in fixed supply and began with the assumption that the resource can be extracted at a constant average and marginal cost. The firm may decide to extract the entire deposit over time, or it may cease extraction before the limit is reached. Moreover, in Gordon's model the firm may decide to extract nothing during the current period when future conditions are particularly favorable.

Given a fixed supply of resources, the firm will choose the rate of output at which the value of the remaining deposit will grow at the firm's internal rate of discount over time. This growth can be produced by a rise in price over time, a fall in production costs, or some combination of the two. If demand or the cost function changes over time, the rate of output may increase or decrease. Gordon also showed that the firm will never offset a substantial shift in either demand or cost that applies to all periods by increasing the rate of output over time. Finally, Gordon extended the model to the industry; this extension will be mentioned in the next subsection.

Theory of the Mineral-Extracting Industry

We will treat the theory of the mineral-extracting industry just as we did that of the mineral-extracting firm. That is, we will summarize only a few of the more important papers on the topic. These few papers will give a general impression of most of the papers in this area.

Let us begin by summarizing the latter portion of the 1955 paper by Herfindahl, the first part of which, the theory of the firm, was summarized in the previous subsection.[8] One of Herfindahl's basic contributions was bringing exploration into the theory of mineral extraction. The first case Herfindahl considered was the one in which both the quantity of discovered ore and the average cost of discovering a unit of ore remain constant. If the industry is characterized by free entry, in long-run equilibrium there will be zero profits from both exploration and mining. All profits on known deposits will be eliminated by the cost of exploration. That is, so long as there is some expected return to exploration, additional exploration will be undertaken. In a competitive industry no property will remain idle unless, of course, costs are so high that variable costs cannot be covered. Thus, after an increase in demand, price, and therefore profits, will increase. This increase in profits encourages exploration and increased extraction until profits again fall to zero. This will be an extremely useful assumption in our later analysis.

In Herfindahl's model the amount of known reserves is determined by market forces. The basic determinant that changes the number of properties and the rate of exploration is profit. Five variables determine the rate of output and exploration.

1. *Demand.* Changes in demand change price and therefore profits. Profits determine exploration and increases in output. Output increases causing price to return to its original level. Both the known stock of the resource and the extent of exploration remain greater.
2. *Exploration Costs.* A change in the cost of exploration changes profit and affects exploration expenditures. When profit is driven back to zero, the output and the stock of properties will be different.
3. *Economies of Scale in Mining.* The larger the range of output over which increasing returns prevail, the more rapidly a given property will be exploited, and the lower the price of ore.
4. *The Rate of Discount.* The higher the rate of discount, the greater the return to current income, and the more rapidly a given property will be exploited.
5. *The Average Size of Property.* If increasing marginal costs exist after some given output, increased property size leads to a longer mine life. By extending the mine life, the owner escapes some diseconomies.

As Herfindahl indicated, these five forces are not the sole determinants of the known deposits and the rate of exploitation from known properties, but they are the principal forces determining these variables.

Herfindahl next assumed that the cost of exploration rises with increased exploration. Again, changes in the number of mines are generated by profits or losses in exploration, that is, by the difference in the value of a typical known property and the cost of discovery. Under increasing rather than constant costs of exploration the life of a typical mine will be greater and the stock of properties lower. Since exploration costs rise with increased exploration, the return to exploration is lower, and there will be less exploration than under constant exploration costs. As a result there will be fewer known properties than under the previous assumptions. As output continues and the cost of exploration rises, the value of known properties increases. The discovery rate and the stock of properties fall. The price of ore rises as exploration becomes more expensive.

Herfindahl went on to consider the Hotelling model in which discovery is impossible. Price in this case rises each year by the rate of discount. Any individual owner is, therefore, indifferent as to when he exploits his property. If discovery is possible but exploration can be carried out only under very rapidly increasing costs, most output is from reduced stock, but as price rises, some exploration and discovery take place. In this case price does not rise at a rate as high as the rate of discount.

Possibly some ore can be imported from other countries or regions, but profits still regulate exploration in each region, even though the costs of exploration are not all the same. In equilibrium the costs of exploration in all regions will rise together. A newly discovered region, if important enough, will obviously lower commodity price and exploration in other regions.

The paper finishes by noting two complications to the analysis. First, technological development may overcome or at least postpone long-run difficulties. Also metals may not be independent in extraction, because metals are frequently found together. Thus, exploration often involves more than one type of ore.

Herfindahl tested his hypothesis that the rate of exploration is related to expected profit in the case of the petroleum industry in the United States. He found a significant positive relation between the two variables.

In a later paper Herfindahl carried the analysis several steps further with a simple case in which all deposits are known and of uniform quality, and demand remains constant from year to year.[9] The royalty—the price of a unit of ore in the ground, which is resource price less production cost—must rise over time at a rate equal to the rate of interest. Assuming a constant average cost of extraction for the industry, the price of the resource must increase over time. Herfindahl assumed a yearly demand for the resource as depicted in panel A, Figure 2.4. He then noted that with a fixed deposit, a given interest rate, and the given demand curve with a maximum price at which any output can be sold, the firm would choose a time path of production at which the last unit of output at exhaustion would be sold at the maximum price and price would rise over time at the same rate as the interest rate. The path of the resource price is shown by AR in Figure 2.4, panel B. Since OC is average cost, the difference between AR and OC is royalty. Thus, current price and production are determined by the magnitude of the deposit, the interest rate, and demand.

FIGURE 2.4

Exhaustion of a Resource

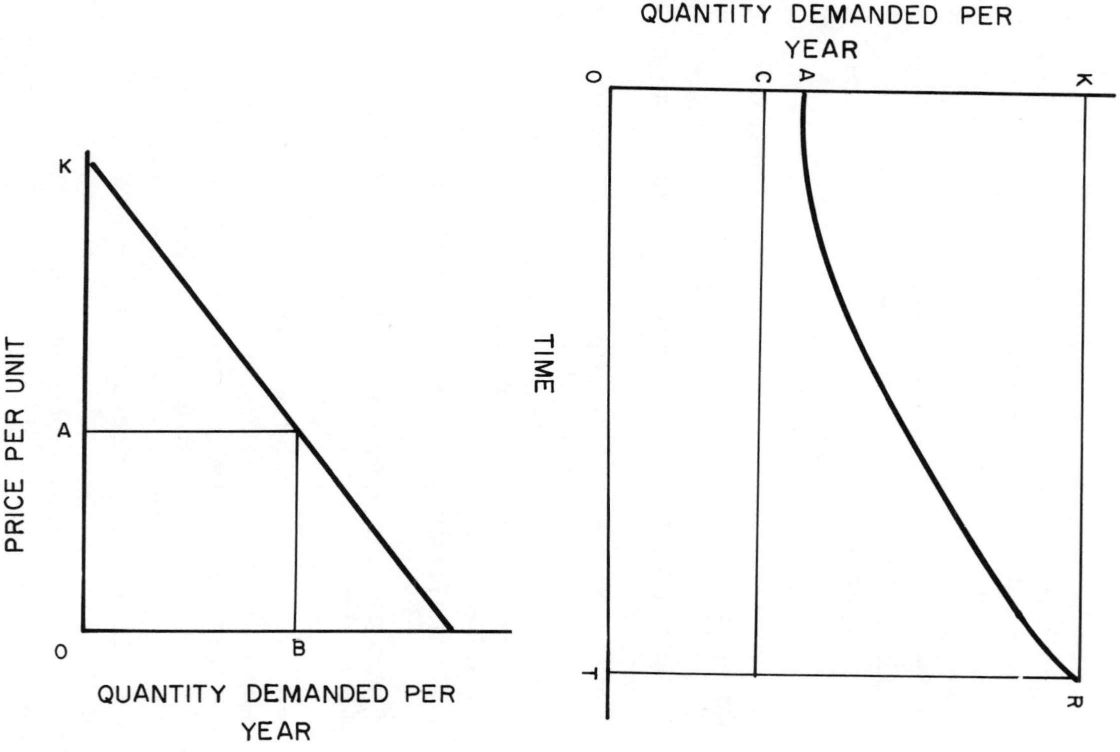

Herfindahl noted that holding a mineral deposit is simply an alternative to an investment in other capital or to consumption. The discount rate allows a comparison. The market will equate the returns to all alternatives at the margin. The mineral deposit is merely another form of capital. It yields a flow of services and can wear out, just like a machine.

The next part of Herfindahl's paper is devoted to several analyses of the effect of changes in some of the parameters of the model. The first change is a reduction in the rate of interest. If the initial resource price is the same, OA in Figure 2.4., the resource price, rises less rapidly at the lower rate of interest; that is, AR pivots downward in panel B. Since the lower price in each period leads to greater sales, the total deposit is exhausted before time T and before price reaches OF. Therefore, the industry solution is to reduce current output, producing a higher current price. The earlier prices are higher, but the later prices are lower, if the price OK is attained at exhaustion. The time horizon is extended beyond T. In other words with a lower interest rate, present extraction is not so valuable, and extraction is postponed extending the life of the deposit.

An increase in the quantity of deposits would lower price, increase quantities in each period, and extend the time horizon. If average cost, OC in Figure 2.4B, falls, the initial price and all early prices must fall. Later prices are increased, and the time horizon until a price of OK is reached is shortened; the initial royalty is increased. An increase in cost has the opposite effect: Initial price is increased, and production shifts to the future.

A severance tax per unit of output has the same effect as an increase in cost. However, Herfindahl noted that a royalty-based severance tax that takes constant proportions of the royalty will not affect prices, quantities, or the time horizon. Note that royalty is price less average cost, or the price of the resource in the ground. This type of tax changes the present value of each royalty but does not change the relation of each pair of royalties. The extraction schedule would not be changed. We might note, however, that this result is obtained only under the very restrictive assumptions of Herfindahl's paper. We will discuss the impact of taxation more completely in Chapter 7.

Herfindahl considered the consequences of a shift in commodity demand. The results were shown to depend upon the way in which demand shifts, particularly on whether or not the maximum price changes. He then modified the model to consider limited quantities of known deposits, of varying quality. He abstracted to consider only two qualities of ore, each physically separated from the other. The price of ore from either grade of deposit is the same, but the cost of extracting from the poorer grade is higher. The better grade of ore will be extracted first. Since profit from the better grade is higher, present value is higher when these higher profits are reaped first and are allowed to grow over a longer horizon through investment, and the lower profit ore is exploited later.

The assumptions of the model were then changed to allow for exploration. That is, all deposits are not known. This change complicates the theory but the basic results still hold. The individual cases will not be summarized here.

Herfindahl concluded that in the mineral industries price behavior is consistent with the view that very large quantities in relation to current consumption will be available for a long time at costs near present costs. That is, the mineral industries have not acted as if they are running out of ore. To be sure, firms extract all that is economically feasible from given deposits, but they expect other deposits and technological change.

As mentioned above, the paper by Richard Gordon dealt not only with the mineral-extracting firm but also the mineral-extracting industry. As Herfindahl also noted, Gordon showed that (with constant production cost) price rises over the period of exhaustion at the interest rate, and price in the terminal period will be that price at which quantity demanded equals zero.[10] Any distribution over time will give the same present value to the firm in this model, but industry totals are fixed and determinant. The conclusion is shown to apply only to extraction under conditions of constant average cost. If firms produce under conditions of increasing cost, the time path of resource prices cannot be easily determined. That is, profits do not necessarily grow at the interest rate. The purely qualitative behavior of prices and output remains similar, however. Gordon concluded from his analysis that there were apparently at that time no industries in which exhaustion was a pressing problem. That is, no industries were behaving consistently with the exhaustion model. Fears of resource wastage, therefore, he noted, might be groundless.

In the late 1960s Vernon L. Smith and Ronald G. Cummings each wrote papers that advanced considerably the general theory of extraction of exhaustible resources.[11] Because these papers are only tangentially related to the specific application of the theory to mining, we will summarize them very briefly here.

Smith developed a general theory that fits the case of extraction from an exhaustible but replenishable resource. His theory also covered the possibility of a common property resource such as petroleum or fish. In the case of mining, in marked contrast to the other cases, very little that had not been previously deduced was added.

Cummings' paper was concerned with the common property problem. He considered two situations: (1) A centralized manager organizes the production of a common property resource with many firms, and (2) one firm owns the resource. He also considered the problem of costs that are affected by cumulative production. Boundary conditions—zero rates of extraction—were analyzed. He concluded that production from a common property resource is extended until profits become zero. If decisions are then assigned to a central authority, it maximizes social benefit by equating among all firms marginal production costs plus area-wide user costs. When costs increase because of cumulative production, the rate of extraction is reduced, the production period extended, and the rate of growth in marginal profits reduced.

In his Richard T. Ely Lecture at the 1973 meeting of the American Economic Association, Robert M. Solow brought together most of the current theories of exhaustible resources.[12] He began with the now familiar observation that the owner of a resource deposit views that deposit simply as a capital asset. The only difference between them is that a capital asset is reproducible whereas the deposit is not. A resource deposit is valuable because it can be extracted and sold. Left in the ground, it produces a return only by appreciating in value. Asset markets can be in equilibrium only when all assets in the same risk class earn the same rate of return, which in this case is the interest rate for that risk class. Therefore, Solow derived the generalized argument that the value of a resource deposit must grow at a rate equal to the market rate of interest. Thus, the price of ore, net of extraction costs must increase exponentially at the same rate as the rate of interest. Under competition, net price is the market price less marginal cost; under monopoly, it is marginal revenue less marginal cost. Of course, this whole argument is similar to that of some earlier papers summarized here. Solow derived the results using an asset argument rather than a flow argument, however.

Mineowners are indifferent in the long run between extracting the resource and holding it. Equilibrium holds in the market at each period. If net price rises too slowly, extraction is accelerated and the resource is exhausted too quickly; if net price rises too fast, owners would delay extraction.

As Solow noted, this analysis does not mean that the market price rises exponentially. The price of ore in the ground, which under competition is market price less extraction cost, rises at a rate equal to the rate of interest. If extraction costs are falling, the market price can fall while net price rises. Thus, as the resource becomes increasingly scarce, the net price rises to reflect this scarcity. Eventually, of course, the market price must rise unless new deposits are located, because of the limit to which extraction costs can fall. As the market price rises, the quantity sold, and hence extracted, must fall, because the demand for the resource is downsloping. Finally, price will become so high that none is demanded. When everything has been coordinated perfectly, this price will be reached the instant that the deposit is exhausted.

Solow used the same tools, assets rather than flows, to analyze the multigrade problem. Not surprisingly he came up with the same solution described above. The better grades—those with lower extraction costs—will be exhausted before extraction from poorer deposits begins. He then showed the possibility that the equilibrium path might not always be stable if the future markets were not perfect. We need not develop these arguments here. Solow concluded that in tranquil times the equilibrium path would be followed rather well, but the resource markets would probably respond rather drastically to unexpected shocks or sudden surprises.

In theory, as Solow noted, the externally set interest rate (set externally to the industry in question) determines the "correct" equilibrium path. But he stressed also that this path is correct only to the extent that the market rate of interest correctly indicates the social rate of time preference. As one would expect, the social rate of time preference is a rather nebulous concept in that it is difficult to imagine a social utility function, particularly when we are dealing with unborn generations. In any case, the choice of the proper discount is rather crucial to the problem. Solow spent a considerable portion of his lecture discussing the concept of a proper rate, but we will not attempt to summarize his arguments here, because our problems involve different aspects of the pure theory of exhaustible resources. We will also pass over his discussion of additional welfare aspects of the theory.

APPENDIX NOTES

1. Frederick M. Peterson and Anthony C. Fisher, "The Exploitation of Extractive Resources: A Survey," *The Economic Journal* (December 1977): 681–721.

2. Lewis C. Gray, "Rent under the Assumption of Exhaustibility," *Quarterly Journal of Economics* (May 1914): 464–89. Reprinted in Mason Gaffney (ed.), *Extractive Resources and Taxation* (Madison: University of Wisconsin Press, 1967) 423–46.

3. Harold Hotelling, "The Economics of Exhaustible Resources," *Journal of Political Economy* (April 1931) 137–75.

4. Donald Carlisle, "The Economics of a Fund Resource with Particular Reference to Mining," *American Economic Review* (September 1954): 595–616.

5. Orris C. Herfindahl, "Some Fundamentals of Mineral Economics," *Land Economics* (May 1955): 131–38.

6. Anthony T. Scott, "The Theory of the Mine under Conditions of Certainty," in Gaffney (ed.), *op. cit.*, 25–62.

7. Richard L. Gordon, "A Reinterpretation of the Pure Theory of Exhaustion," *Journal of Political Economy* (June 1967): 274–86.

8. Herfindahl, *op. cit.*

9. Orris C. Herfindahl, "Depletion and Economic Theory," in Mason Gaffney (ed.), *op. cit.*, 63–90.

10. Gordon, *op. cit.*

11. Vernon L. Smith, "Economics of Production from Natural Resources," *American Economic Review* (June 1968): 409–31; Ronald L. Cummings, "Some Extensions of the Economics Theory of Exhaustible Resources," *Western Economic Journal* (September 1969): 201–10.

12. Robert M. Solow, "The Economics of Resources or the Resources of Economics," *American Economic Review* (May 1974): 1–14.

3

The Competitiveness of Mineral-extracting Industries: Some Empirical Evidence

In Chapter 2 we emphasized that a significant feature of the theoretical model of the competitive industry is that the forces of competition tend to eliminate above-normal and below-normal profits for firms in the long run through changes in investment, production, and market price. Since we will employ a long-run competitive model in most of our empirical estimates, we must discuss some of the important implications of this model and test to see how well the mineral extracting industry fits our theory. After a discussion of the theoretical implications, we will first use data from Canadian corporations and then use data from mining companies traded on financial markets in the United States to see how well the competitive theory applies to the mineral-extracting sector.

It will be shown that during the period for which data are available, rates of return in the Canadian mining industry did not differ significantly from the prevailing rate of return in the economy as a whole; thus, in the case of Canada the mining industry seems to fit the theory quite well. When evidence from firms traded on financial markets in the United States is examined, the data show that in most cases, the returns in mining, do not differ significantly from the going rate in the economy as a whole. It follows then that the results for both the Canadian mining industry and the mining firms whose stocks are traded on the equities market in the United States suggest the presence of competition, which through the free flow of investment capital tends to allocate resources to eliminate abnormal profits through price and output adjustments.

SOME IMPLICATIONS OF THE MODEL

We showed in Chapter 2 the way in which the forces of competition tend to eliminate abnormal rates of return on capital. If the rate of return on capital, after taxation, in one or more sectors of the economy is significantly above the going rate of return obtainable in the economy as a whole (abstracting presently from risk differences), investment is attracted from the lower-return sectors into the higher-return sectors. The shift in investment continues until the rate of return in the sector experiencing increasing investment in driven to that in the economy as a whole.

This drive towards equilibrium is accomplished in several ways. First, to the extent that the output of the expanding sector affects the market price, the increased supply resulting from the expansion in the high-return sector drives down the price. The decreased price leads to decreased rates of return. Second, the expansion in the sector receiving investment can cause the prices of certain factors of production or resources used in that sector to be forced upward. The increased prices of resources bring about higher costs, and higher costs drive down the rate of return. Finally, as the expenditure for investment is expanded, investments with lower rates of return will be undertaken, therefore, driving down the average rate of return in the sector. For example, if the rate of return in, say, iron mining suddenly became much higher than the rates in other sectors of the economy, possibly because of an increase in the demand for steel, increased investment would be attracted into iron. If all of the most productive orebodies are already being worked, this new investment would go into less productive areas. Thus, the average rate of return would be driven to the going rate for the economy.

Of course, as noted above, this analysis ignores to some extent differences in risk. If one sector of the economy involves higher risks than others, the return from *successful* investment will be higher in that sector to compensate investors for the higher probability of failure. Of course, the expected rates for *all* investment will be driven down to the going rate in the long run, if investors are relatively risk neutral. To the extent that investors are averse to risk, the going rate in the riskier sector will be somewhat higher.

We might stress that the pattern described above is a long-run rather than a short-run phenomenon of a reasonably competitive economy. Certainly after-tax rates of return can and generally will differ significantly among different sectors of the economy at any specific time or during a particular year or even over several years. These differences can be the results of such things as exogenous shifts in demands, technological changes, or alterations in governmental institutions. All our analysis says is that when the after-tax rate of return in one sector becomes abnormally high, investment is attracted into that sector, and there is a tendency for the relative rates of return to be driven down. When the rates in a sector are abnormally below those in the economy, there is a tendency for capital to leave that sector; the relative rate is then forced upward.

Therefore, even though the rate of return in one sector of the economy differs from year to year from the rate of return for the economy, as a whole there should be no significant difference in the rates over a reasonably long period of time. If the rates of return in that sector remain significantly higher than the going rates, there is some evidence, though not conclusive, that collusion or monopolistic elements, whether governmentally imposed or otherwise, have protected the higher rates by limiting in some way the increase in investment that would have driven down the return.

We might note that there is not the same tendency for before-tax returns to be driven toward equality. It is, after all, the net returns after taxation in which investors are interested. We should also reemphasize that these results are long-run phenomena. Differences may exist over long periods of time because investment is not instantaneously responsive to exogenous changes. This is probably quite true in mining, a sector of the economy in which the returns from an investment frequently lag substantially behind the initial period of investment. Even so, if the industry is reasonably competitive, the after-tax returns will not differ significantly from the rate for the economy over a reasonably long period of time.

The remainder of this chapter presents the statistical data and discusses the tests. Since all tests indicate that the average yearly rate of return in mining does not differ significantly from that in the economy as a whole, we feel justified in using a competitive model for our analyses set forth in later chapters. The reader uninterested in the actual statistics can omit the remainder of this chapter without loss of continuity.

SOME EVIDENCE FROM CANADIAN INDUSTRIES

To examine rates of return in Canadian mining industries and compare these with the social rate of return, we chose two ratios to indicate the return to capital. The first is the ratio of profits after taxation to capital employed. The second is the ratio of profits after taxation to total equity.* Both these ratios should reflect the relevant rate of return. We used both ratios simply as a check

---

*These ratios are the same as those calculated in Appendix B of *Statistics Canada* for 1970. Following the formula used there, we calculated the first rate of return as the ratio of after-tax profits to total liabilities and equity minus total current liabilities. For the second rate we used their formula, the ratio of after-tax profits to total equity plus long-term amounts due to shareholders or affiliates.

for the consistency of our results. Other ratios, of course, can be computed, but these are generally not as useful for comparative purposes.

We considered the period 1962–75. For the rate of return on investment for the total economy, we used the rate of return in total manufacturing in each year. This rate seems to reflect best the true opportunity cost of capital investment in other sectors. To determine the competitiveness of the mineral industry, we compare this rate of return with the rates of return in mining in the Canadian economy.

Table 3.1 shows the relevant rates of return that were calculated. Columns (2) and (3) list the returns on capital employed in total manufacturing and in mining respectively for each of the 14 years. Columns (4) and (5) show the returns on total equity in each year for the two sectors of the economy. It is easily seen upon casual observation that the returns to mining and manufacturing were in general rather close over the period under consideration. Sometimes mining had a bit higher return; sometimes manufacturing was a little higher.

TABLE 3.1

Rates of Return in Canada: Manufacturing and Mining, after Taxation

Year (1)	Profits to Capital Employed		Profits to Equity Employed	
	Manufacturing (2)	Mining (3)	Manufacturing (4)	Mining (5)
1962	0.064	0.070	0.079	0.083
1963	0.073	0.073	0.089	0.087
1964	0.078	0.094	0.097	0.110
1965	0.077	0.096	0.097	0.113
1966	0.074	0.081	0.095	0.096
1967	0.061	0.085	0.079	0.101
1968	0.066	0.082	0.085	0.100
1969	0.068	0.072	0.088	0.088
1970	0.048	0.063	0.062	0.078
1971	0.059	0.047	0.077	0.060
1972	0.073	0.048	0.099	0.065
1973	0.097	0.097	0.128	0.129
1974	0.113	0.102	0.150	0.138
1975	0.090	0.086	0.121	0.119

Source: Ratios calculated from yearly statistics reported by *Statistics Canada*.

Figures 3.1 and 3.2 show graphically the closeness of the two rates of return over the years. Figure 3.1 presents a graph of the rates of return to capital in manufacturing and in mining; Figure 3.2 shows the after-tax returns to total equity in the two sectors. The two paths are by and large rather similar. In both cases the returns are obviously quite close.

We can test the closeness of the two rates more precisely by statistically testing whether the mean difference between the rates of return in the two sectors is statistically different from zero. That is, we test the hypothesis that the average difference between the yearly rates of return in manufacturing and in mining is zero. Table 3.2 shows the yearly differences in returns over the time period. Column (2) lists the profit-to-capital ratio in manufacturing minus the same ratio in mining for each year. The average of these differences is a very low -0.003929; the standard deviation is 0.01378. Testing the hypothesis that the true average difference is zero, we get a t-value of 1.067.* The t-value is too small at any relevant level of significance to reject the hypothesis that the average difference is zero. Thus we must accept the hypothesis that there is no significant difference in these rates of returns.

Column (3) shows the yearly differences between rates of return on equity in manufacturing and in mining. As noted at the bottom of the column, the average difference is an even smaller: -0.0015, with a standard deviation of 0.015133. With a t-value of 0.3709 we must at any relevant level of significance clearly be unable to reject the hypothesis that the average difference is zero. In each case, using the relevant rate of return on capital, our results indicate that the rate of return on capital in mining is not significantly different from the going rate on capital in manufacturing as a whole, at least over the only period for which we have meaningful statistics: 1962–75. For this reason we have a certain amount of evidence to justify our use of a theoretical model in which rates of return are driven toward the going rate in the economy over the long run.

This is, of course, not to say that changes in the tax laws have not and will not have a significant impact upon rates of return. In fact, that impact is one of the most important things that we wish to analyse. But the reader should note that this is a long-run model. Since investment in mining is a long-term process, probably the full impact has not as yet been felt. It simply takes time for firms to adjust to exogenous changes when considerable uncertainty prevails. Our

---

*If this test we are assuming that our data represent a random sample drawn from a normally distributed population, $x_i \sim N(\mu, \sigma^2/n)$. Hence, the mean will also be distributed normally, $\bar{x} \sim N(\mu, \sigma^2/n)$ where n is the number of observations; and, since we must estimate the variance, the appropriate test statistic is a t calculated as $\bar{x}/(s/\sqrt{n})$, where s is the estimated standard deviation.

FIGURE 3.1

Profits to Capital Employed after Taxation (Canada)

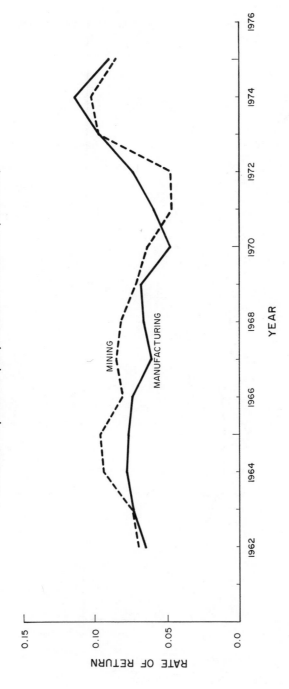

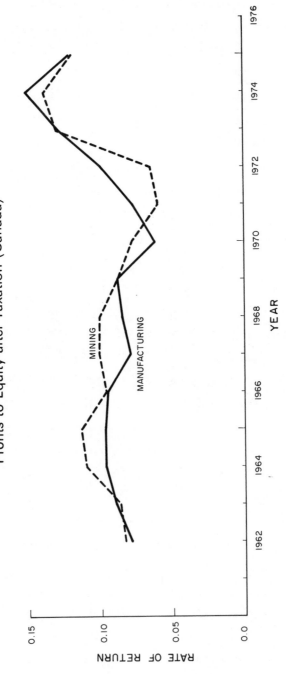

FIGURE 3.2

Profits to Equity after Taxation (Canada)

56

TABLE 3.2

Yearly Differences in Rates of Return in Canada: Total Manufacturing Minus Total Mining, After Taxation

Year (1)	Difference in Profits-to-Capital-Employed Ratios (2)	Difference in Profits-to-Equity Ratios (3)
1962	−0.006	−0.004
1963	0.000	0.002
1964	−0.016	−0.013
1965	−0.019	−0.016
1966	−0.007	−0.001
1967	−0.024	−0.022
1968	−0.016	−0.015
1969	−0.004	0.000
1970	−0.015	−0.016
1971	0.012	0.017
1972	0.025	0.034
1973	0.000	−0.001
1974	0.011	0.012
1975	0.004	0.002
Average	−0.003929	−0.0015
Standard deviation	0.013784	0.015133
t-value	1.067	0.3709

results were really derived under conditions in which the tax laws were relatively stable. In any case, we feel justified empirically in using the model in which rates of return are equalized in the long run. We will return to the effect of tax changes in Chapter 7.

SOME EVIDENCE FROM MINING INDUSTRIES TRADED IN FINANCIAL MARKETS IN THE UNITED STATES

We will now analyze profitability data for mineral-extracting firms traded in U.S. financial markets. These data, from Standard and Poors, *Analyst's Handbook*, offer a useful source of corollary information.* Available data from

---

*These were the only relevant returns available for U.S. industries over the period. They should be rather comparable to the Canadian data presented above.

Standard and Poors go back to 1951, a full decade before Canadian data were available. Rates of return in groups of firms traded in U.S. financial markets probably are somewhat closely related to rates of return in Canadian industries because of the relatively free flow of capital between the two countries. If rates of return were significantly higher in one country, investment capital would in general move over time from the country with the lower return to the country with the higher. The two rates under these circumstances would tend to equalize in the long run.

TABLE 3.3

Rates of Return in United States after Taxation and Depreciation (Profit to Book Value)

Year (1)	425 Industrials (2)	Miscellaneous Mining (3)	Copper (4)	Lead and Zinc (5)
1951	0.131	0.164	0.117	0.139
1952	0.117	0.123	0.096	0.094
1953	0.121	0.089	0.098	0.045
1954	0.115	0.105	0.088	0.060
1955	0.137	0.149	0.164	0.097
1956	0.128	0.142	0.180	0.099
1957	0.119	0.108	0.074	0.123
1958	0.095	0.056	0.058	0.050
1959	0.105	0.097	0.065	0.078
1960	0.098	0.100	0.073	0.075
1961	0.094	0.096	0.066	0.065
1962	0.103	0.090	0.071	0.053
1963	0.109	0.112	0.064	0.104
1964	0.118	0.140	0.077	0.128
1965	0.125	0.137	0.100	0.161
1966	0.137	0.115	0.123	0.146
1967	0.124	0.104	0.077	0.124
1968	0.127	0.102	0.092	0.139
1969	0.128	0.086	0.146	0.169
1970	0.117	0.122	0.112	0.107
1971	0.123	0.073	0.066	0.050
1972	0.131	0.096	0.081	0.104
1973	0.156	0.139	0.114	0.198
1974	0.161	0.174	0.121	0.223

Source: Standard and Poors, *Analyst's Handbook*, 1975.

Table 3.3 shows rates of return from 1951 through 1974 for (1) the Standard and Poors Index of 425 Industrials, (2) miscellaneous mining, (3) copper, and (4) lead and zinc.* These returns are computed from data in the 1975 *Analyst's Handbook*. The rates of return are expressed as the ratios of profit after taxation and after depreciation to book value. According to the definition used by Standard and Poors, book value consists of a total of common stock, capital surplus, and retained earnings, less treasury stock, intangibles, and the difference between the carrying value and liquidating value of preferred stock. While this ratio is not the same as the profit-to-capital ratio used for Canadian industries, it should give some indication of the return to capital. But we must emphasize that the ratios may differ somewhat, because the value of capital employed differs from book value as defined above.

In any case, using the ratio of profit to book value as the after-tax rate of return, one can see from Table 3.3 that the returns in miscellaneous mining, copper, and lead and zinc were not greatly different from the rate of return in the index of 425 Industrials, which we use as a proxy for the going rate of return in the economy as a whole. Figure 3.3 illustrates much more clearly the generally close relation between the rates in each of the three segments of the mining industry and the rate for the 425 Industrials.

Table 3.4 gives the yearly differences between the rate of return in each mining sector and that for the index of 425 Industries; column (2) shows miscellaneous mining; column (3) shows copper; and column (4) shows lead and zinc. In all three cases the average difference was positive; that is, the 425 Industrials had, on the average, higher returns, but the average differences were quite small. The average of 425 Industrials minus miscellaneous mining was only about 0.008, and the 425 Industrials minus lead and zinc only 0.012. Neither of these averages was significantly different from zero at the 5-percent level of significance. Thus, we would accept the hypothesis of no significant difference.

The average of 425 Industrials minus copper was about 0.025, not a particularly high figure but still significantly different from zero at the 5-percent level. This result suggests that investment in copper was on average carried out to such extent that returns were driven below the going rate for 425 Industrials. Whether the copper industry in the United States was suffering the impact of a disturbance to which it had not adjusted over the period of analysis

---

*While data for gold mining were also available, we omitted rates of return in that industry, because they were substantially below the rates for 425 Industrials in all years. One obvious reason for this discrepancy over such a long period is that during most of the period the price of gold was artificially maintained by the U.S. government. Generally, the prices of other metals were allowed to fluctuate.

FIGURE 3.3

Rate of Profit to Book Value after Taxation (United States)

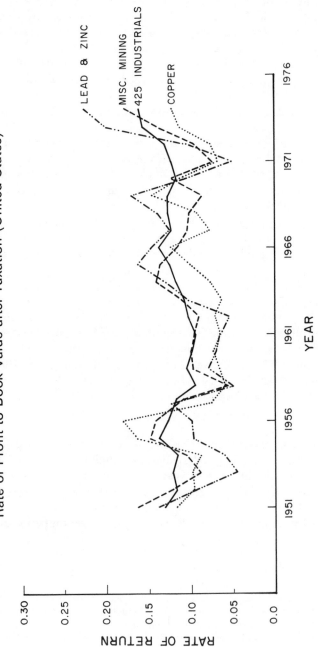

TABLE 3.4

Yearly Differences in Rates of Return in United States after Taxation and Depreciation
(Profits to Book Value)

Year (1)	425 Industrials Minus Miscellaneous Mining (2)	425 Industrials Minus Copper (3)	425 Industrials Minus Lead and Zinc (4)
1951	− 0.033	0.014	− 0.008
1952	− 0.006	0.021	0.023
1953	0.032	0.023	0.076
1954	0.010	0.027	0.055
1955	− 0.012	− 0.027	0.040
1956	− 0.014	− 0.052	0.029
1957	0.011	0.045	− 0.004
1958	0.039	0.037	0.045
1959	0.008	0.040	0.027
1960	− 0.002	0.025	0.023
1961	− 0.002	0.028	0.029
1962	0.013	0.032	0.050
1963	− 0.003	0.045	0.005
1964	− 0.022	0.041	− 0.010
1965	− 0.012	0.025	− 0.036
1966	0.022	0.014	0.009
1967	0.020	0.047	0.000
1968	0.025	0.035	− 0.012
1969	0.042	− 0.018	− 0.041
1970	− 0.005	0.005	0.010
1971	0.050	0.057	0.073
1972	0.035	0.050	0.027
1973	0.017	0.042	− 0.042
1974	− 0.013	0.040	− 0.062
Average	0.00833	0.02483	0.01200
Standard deviation	0.02168	0.02567	0.03605
t-values	1.8823	4.7386	1.6304

is beyond the scope of this study. In general, however, the results of Table 3.4 strongly suggest the presence of market forces tending over time to equate the rate of return in mining to the average rate of return available in the economy.* Thus, our findings for firms traded on the U.S. equity market are in general consistent with our results for Canadian Mining.

CONCLUSION

It appears from the evidence presented in this chapter that we are justified in using a model that assumes investment in mineral-extracting industries to be undertaken until the long-run rate of return equals the rate of return in alternate investments—that is, the going rate of return to capital in the economy. In no case for any mining sector examined did we find evidence that, over the long run, the returns to capital in mining exceeded the returns in all manufacturing or in all industrials, either for the United States or for Canada. Indeed, our evidence would indicate that the rate of return to capital in the mineral extraction sector is equal to that in the total economy and therefore supports our assumption of a competitive industry.

---

*It may be the case that the 425 Industrials, being composed of larger firms, could have slightly higher returns than the true going rate in the economy.

4

The Production Relation in
Mineral Extraction:
Some Empirical Evidence

As shown in Chapter 2, the cost of producing a particular amount of output depends in large measure on the form of the production function. Thus, to examine an industry or sector of the economy, one needs certain information about the production relations in that industry or sector. In this chapter we will present some empirical evidence concerning the specific form of the production function in the mineral-extraction sector of the economy. In this way we can gain considerable insight into the way production is carried out in mineral production.

We shall focus our attention specifically on three major features of the mineral-production relations. First, we examine whether changes in the level of output or scale of operation change relative usage of inputs. Clearly, producing more output requires using more inputs. However, we are interested in whether or not the *ratios* of input usage change as the level of output changes.

In more technical terms we will examine the mineral-production function to see whether it is homothetic or nonhomothetic. A production function is said to be homothetic if changes in the level of output have no effect on the

This chapter is based on the results of a study performed by the authors for the Ministry of Natural Resources, Ontario, Canada. (Charles W. Smithson, Gerhard Anders, W. Philip Gramm, and S. Charles Maurice, *Factor Substitution and Biased Technical Change in the Canadian Mining Industry* [Ministry of Natural Resources: Ontario], 1979.)

relative usage levels or ratios of inputs. If the production function is nonhomothetic, an increase in the rate of extraction would cause the mineral-extraction sector to use more of some inputs relative to others, independent of any change in the relative prices of the inputs.

The issue of homothetic versus nonhomothetic mineral-production functions is important for two reasons. First, it is important to learn whether the mineral extraction sector will use some inputs more or less intensively as that sector expands. Second, since several previous studies of input substitutability statistically imposed the condition of a homothetic production function, the actual existence or nonexistence of this condition is important when comparing our results with previously derived results for other sectors.

The second major point examined in this chapter is the effect of technological change in the mineral-extraction sector on relative input usage in that sector. Technological change can result either from innovation or from an input price-induced change from one production technology to an alternative form.

In technical terms, technological change of either type is said to be neutral if, for any level of output, the ratios or relative usages of inputs is unchanged —even though the absolute amount of inputs are decreased for each level. Technological change is said to be biased if it changes the ratios or relative usage of inputs. For example, if there are only two inputs used in production (capital and labor), technological change is neutral when the capital-labor ratio remains unchanged after the change in technology. If the change in technology causes an increase in the usage of labor relative to capital for a given level of output, technological change is said to be *labor using* or *capital saving*. If the capital-labor ratio increases, the bias is toward using capital or saving labor. Note that it is the change in the *ratio* for a given level of output and not the change in the absolute level of usages that determines neutrality or the direction of bias. The presence and direction of bias are important points to be considered when examining production relation in the mineral-extracting sector of the economy and evaluating policy changes.

Finally, and perhaps most important, we shall examine the production relations to determine the extent to which inputs can be substituted for one another in response to changes in the relative prices of inputs in the mineral-extracting industry. Moreover, when we find that inputs can be substituted for one another, we will also attempt to estimate the degree of substitutability among inputs used in mineral extraction.

The existence and extent of substitutability in the mineral sectors is extremely important now because of the so-called doomsday models that predict exhaustion of the stock of many natural resources. If, however, the raw natural-resources input can be made less essential in the production process— i.e., if other inputs can be substituted for it—when its price rises in response to increased scarcity, such dire predictions would not be realized. In other words, factor substitution is clearly one way a depletable resource may become less essential. As the stock of the resource declines, its price will increase. Therefore, depending on the possibility and degree of substitution, we would expect the rate of consumption of the depletable resource to decline.

In our analysis we will examine these features empirically, using data from the mining industry in Canada for the period 1962–74. Since the Canadian mining industry is well developed and accounts for a substantial portion of total world output of minerals, we feel that it may be treated as representative in the sense that it would employ the most efficient production technology that is currently available; thus, our results are probably quite general. Furthermore, Canadian data were readily available and quite accurate. For the majority of the analysis we will concentrate on the metal-mining sector, but, we will also present estimates from nonmetal mining.

Since the issue of input substitutability has received considerable attention recently, we begin with a brief review of some preceding empirical evidence from other producing sectors. The next section provides an overview of the Canadian metal mining industry in the period under consideration. The model to be used is developed, and the empirical results concerning homothetic or nonhomothetic production functions, biased or nonbiased technological change, and input substitutability or nonsubstitutability in metal mining are set forth in later sections. We then discuss briefly some results for nonmetal mining and summarize our conclusions.

We must note that the body of this chapter is the most technical of any of the chapters in the book in an econometric and statistical sense. Therefore, the reader who is not interested in or familiar with the econometric techniques used should skip the body of this chapter, which is designed primarily for the professional economist or student of economics. The results of the study, which should be of interest to both economists and noneconomists, are presented in the concluding section of the chapter.

FACTOR SUBSTITUTION: A SURVEY OF THE EVIDENCE

As noted above, the possibility of input substitution has received a considerable amount of attention recently in the economics literature. To date, several studies have considered the issue of factor substitution. These studies concentrated primarily on the substitutability between the energy input and other reproducible factors of production. Our analysis would appear to be one of the first to consider also a depletable resource ore.

Several studies have considered the substitutability between capital, labor, energy, and nonenergy intermediate inputs in the manufacturing sector. Ernst R. Berndt and David O. Wood used time-series data drawn from U.S. manufacturing, 1947–71, and a methodology similar to the type we shall employ.[1] Their estimates indicated substitutability between capital and labor, energy and labor, and intermediate nonenergy inputs with all of the other factors of production and complementarity (nonsubstitutability) between capital and energy. Similar econometric findings were reported by Berndt and D.W. Jorgenson, using slightly different industrial data for the same time period.[2] M.A. Fuss used pooled cross-section data drawn from Canadian manufacturing, 1961–71, and the same methodology.[3] With the exception of a finding of complementarity between energy and intermediate, nonenergy materials, his results were the same as those mentioned above. M. Denny, J.D. May, and C. Pinto used a different methodology to examine time-series data for Canadian manufacturing, 1949–70.[4] In their analysis they tested and were able to reject the hypothesis that the production function is homothetic. They did not, however, explicitly consider the possibility of biased technical change. Their results were the same as those obtained by Berndt and Wood, except that Denny et al. found capital and intermediate materials to be complements rather than substitutes. Of particular significance to our analysis, they found that the estimates were sensitive to the imposition of homotheticity. If homotheticity is imposed on the production relation, the inputs appeared to be significantly less substitutable. Indeed, with homotheticity imposed, labor and energy became complements rather than substitutes.

James M. Griffin and Paul R. Gregory examined the substitutability between inputs in a three-input setting (i.e., capital, labor, and energy), using international cross-section and time-series data covering the period 1955–69 and the methodology employed by Berndt and Wood.[5] As did the preceding four-input studies, Griffin and Gregory found capital and labor and labor and energy to be substitutes, but their estimates indicated that capital and energy are substitutes rather than complements.[6]

Finally, Laurits R. Christensen and William H. Greene considered the question of factor substitutability in a single industry.[7] Using data from U.S. electric utility firms in 1955 and 1970, they estimated the degree of substitutability among capital, labor, and energy used in the generation of electric power. They used a methodology similar to that employed by Berndt and Wood, however, they tested and rejected the hypothesis that the production relation is homothetic. The results of the estimation showed substitutability between capital and labor, capital and energy, and labor and energy.

AN OVERVIEW OF THE METAL-MINING INDUSTRY IN CANADA

In the case of the Canadian metal-mining industry, data availability limited consideration of the inputs employed to capital, labor, electric energy, and raw ore. In order to obtain a homogeneous measure, the output of this

TABLE 4.1

Historical Trends in Canadian Metal Mining

	1962	1974	Percent Change
Input Usage			
Capital	3255.54	7654.03	+ 135.11
(million 1971 dollars)			
Labor	58.243	70.038	+ 20.25
(thousand workers)			
Energy	3.373	10.282	+ 204.83
(trillion KWHs)			
Ore	114.263	307.237	+ 168.89
(million tons)			
Input Prices			
Capital	0.1012	0.1279	+ 26.38
(percent)			
Labor	5.2539	12.3152	+ 134.40
(thousand dollars per worker)			
Energy	0.7003	0.7554	+ 7.87
(cents per KWH)			
Ore	0.403	2.690	+ 567.49
(dollars per ton)			
Output	579.564	937.508	+ 61.76
(million 1935–39 dollars)			

Source: C. W. Smithson, G. Anders, W. P. Gramm, and S. C. Maurice, *Factor Substitution and Based Technical Change in the Canadian Mining Industry*, Toronto: Ontario Ministry of National Resources, 1975.

sector, ore concentrate, is expressed in constant dollars. Table 4.1 presents some illustrative data on the industry for the period under consideration.[8]

To consider relative input usage, it was necessary that the inputs be expressed in real (physical) quantities. Because labor, energy, and ore were already expressed in physical units, a real capital series was calculated. In this series, the real stock of capital used in the industry during a period was determined by its current and depreciated past real gross investment.*

---

*Specifically, the industry's stock of real capital in period t was defined to be

$$K_t^R = \sum_{i=0}^{n-1} \frac{(1-i\delta)I_{t-i}^G}{P_{t-i}^K}$$

Examination of Table 4.1 indicates that the industry substantially reduced its usage of labor relative to the other inputs. The capital-labor, energy-labor, and ore-labor ratios increased by 95.50 percent, 153.53 percent, and 123.59 percent, respectively. There was also a modest reduction in the usage of capital relative to energy, indicated by a decline of 22.89 percent in the capital-energy ratio. The usage of ore relative to capital and energy remained relatively unchanged (i.e., the capital-ore and energy-ore ratios declined by 12.56 percent and 13.39 percent, respectively).

To obtain comparable input prices, it was necessary to specify a flow price for the services of the capital stock. In this analysis the approach of Jorgenson and Griliches[9] was used to specify the price of capital services as the product of the unit acquisition cost and the user cost (which is itself made up of the real rate of interest and the rate of depreciation).[†] Since in Canada the provincial

---

where I_t^G is gross investment in period t, δ is the straight line rate of depreciation, n is the lifetime of capital, and P_t^K is the price index for capital equipment in period t. In this study the average lifetime of the capital was assumed to be 12 years.

[†]The price of capital services in period t was defined to be

$$P_{K,t} = q_{K,t}(r_t + \delta)$$

where $q_{K,t}$ is the unit acquisition cost of the stock held in period t, r_t is the real rate of interest, and δ is the straight line rate of depreciation (again taking the average lifetime of the capital stock to be 12 years). The nominal capital stock held in t may be expressed as

$$K_t^N = \sum_{i=0}^{n-1} (1 - i\delta)I_{t-i}^G$$

where I_t^G is gross investment in period t and n is the lifetime of capital.

The unit acquisition cost may then be specified to be

$$q_{K,t} = \sum_{i=0}^{n-1} \left[\frac{(1 - i\delta)I_{t-i}^G}{K_t^N} \right] P_{t-i}^K$$

where P_t^K is the price index for capital in period t. Following Lahiri (K. Lahiri, "Inflationary Expectations: Their Formation and Interest Rate Effects," *American Economic Review* [March 1976]: 124–31), a real interest-rate series was estimated. Noting that the nominal interest rate is equal to the sum of the real interest rate and the expected rate of future inflation,

$$i_t = r_t + \dot{P}_t^e,$$

it was assumed that (1) the real rate of interest, is equal to some long-run equilibrium rate with randomly distributed disturbances,

$$r_t = \rho + \mu_t,$$

and (2) the expectations of future inflation are formed by past rates of inflation and the current rate of growth of the money stock,

$$\dot{P}_t^e = \beta_0 \dot{P}_{t-1} + \beta_1 \dot{P}_{t-2} + \ldots + \gamma \dot{M}_t,$$

governments own the mineral resources, the expenditures on ore are the specific mining taxes paid to the provinces, thus the price of ore was specified to be the tax rate per ton.[‡] The most significant increase was the price of labor. The apparent massive increase in the price of ore must be discounted since the entire increase took place in 1973 and 1974. Further, the increase in the price of energy is also misleading since it occurred in the final year. Indeed, over the first half of the period, the price of energy declined.

It would appear then that the significantly reduced usage of labor relative to the other inputs and the modest reduction in capital usage relative to energy could be attributed to neoclassical substitution in response to changes in relative prices. However, it must be noted that, over this period, there was a significant increase in the output of the industry as measured in constant dollar value. If the production function is not homothetic, this output expansion could itself be partially responsible for the changes in relative factor usage. Furthermore, it is quite likely that, over the period under consideration, the industry experienced technical change. If this technical change was nonneutral, it may also have been responsible for changes in relative input usage.

In order to consider explicitly the effect of technical change, this study incorporates an index of exogenous, embodied technical change. Assuming that the technical change is embodied in capital equipment, the index is given by the vintage of the industry's average unit of capital.[*] It should be noted that,

---

where the distribution of past rates of inflation is an infinite geometrically declining series. The resulting equation.

$$i_t = (1 - \lambda)\rho + \beta_0 \dot{P}_{t-1} + \gamma \dot{M}_t - \gamma\lambda \dot{M}_{t-1} + \lambda i_{t-1} + \varepsilon_t,$$

was used to obtain estimates of the long-run real interest rate and a predicted nominal interest-rate series. These were employed to obtain estimates of the expected rate of inflation,

$$\hat{P}_t^e = \hat{i}_t - \hat{\rho},$$

which was then used to estimate a real interest-rate series,

$$\hat{r}_t = i_t - \hat{p}_t^e.$$

[‡]Data on the mining tax alone was not available for Canada as a whole, so the mining-tax data from the province of Ontario was used (this province accounts for more that 40 percent of the metallic mineral output of the nation). The mining tax was divided by tons of ore hoisted to determine the tax rate per ton. This formulation does not account for the "acquisition cost" of the natural resource (i.e., exploration and development costs), but since a majority of these expenditures would be on capital equipment, they are incorporated in the capital variable.

[*]Specifically, this index of technology was calculated as

$$T_t = \sum_{i=0}^{n-1} \left[\frac{(1 - i\delta)I_{t-i}^G}{K_t^N} \right] (t - i)$$

where I_t^G is gross investment in period t, n is the lifetime of capital (assumed to be 12 years), δ is the straight line rate of depreciation, and K_t^N is the nominal capital stock in period t.

while such an index would reflect technological progress in the sense of innovation, it would also reflect technical change in the sense of the applications of known alternative production technologies. Hence, this index might best be viewed as a reflection of the selection among the different subfunctions of production discussed by Furubotn.[10]

THE MODEL

This section sets forth in detail the basic econometric model to be used in carrying out our estimates. While we follow well-known econometric procedures, we should emphasize that the derivation of the model and the model itself are extremely complex in an econometric sense. We shall assume throughout the discussion that the reader is familiar with basic econometric techniques and terminology. As noted, the results are set forth in the concluding section of the chapter for those not interested in technical derivations.

The output of the mining industry, ore concentrate (Q), is produced using capital (K), labor (L), energy (E), raw natural resources (N), and other inputs. Technology (T) enters the production function because it can increase total output by increasing the marginal productivity of the inputs. If weak function separability is assumed in order to eliminate the relatively minor inputs, the net output function may be specified as*

$$Q = f(K, L, E, N; T). \qquad (4.1)$$

While the degree of substitutability between pairs of inputs (i.e., the elasticities of substitution) and the bias in technical change may be estimated from the production function itself, there is the problem that the use of a functional form of the production function capable of providing a second-order approximation to an arbitrary production function fails to yield input demand functions that are explicit functions in the parameters of the production function. Hence, the cost function dual to equation (4.1) will be employed. That is,

$$C = g(P_K, P_L, P_E, P_N, T, Q) \qquad (4.2)$$

---

*Since the purpose of this chapter is to examine some of the neoclassical properties of the production relations, we use a simple one-period production function rather than the dynamic production relationship described in Chapter 2.

where $P_i (i = K, L, E, N)$ denotes input prices, which are exogenous to the firm. It is assumed that (1) the cost function is a positive, continuous function in Q and P_i and approaches infinity as a limit as Q approaches infinity, (2) the cost function is concave and linearly homogeneous in p_i, and (3) the cost function is strictly decreasing with respect to T.

For purposes of estimation it is necessary to specify the cost function in parametric form. The form employed in this analysis is the transcendental logarithmic (translog) form developed by Christensen, Jorgenson, and Lau.[11] This functional form has three major attributes:

1. It may be regarded as a general, second-order log-quadratic local approximation to an arbitrary cost function.
2. While it permits estimation and tests of significance of elasticities of substitution, it entails no *a priori* restriction on their values or constancy.
3. It allows for estimation of additional properties of the underlying production function including homotheticity and the character of technical change.

In translog form, the cost function (2) becomes

$$
\begin{aligned}
\ln C = {} & \alpha_0 + \beta_Q \ln Q + 1/2\beta_{QQ}(\ln Q)^2 \\
& + \phi_T \ln T + 1/2\phi_{TT}(\ln T)^2 + \gamma_{QT}\ln Q \ln T \\
& + \sum_i \gamma_i \ln P_i + 1/2 \sum_i \sum_j \gamma_{ij} \ln P_i \ln P_j \\
& + \sum_i \beta_i \ln Q \ln P_i + \sum_i \phi_i \ln T \ln P_i \quad (i, j = K, L, E, N)
\end{aligned} \quad (4.3)
$$

Linear homogeneity in factor prices requires

$$
\sum_i \alpha_i = 1 \tag{4.4}
$$

$$
\sum_i \gamma_{ij} = \sum_j \gamma_{ji} = \Sigma_i \Sigma_j \gamma_{ij} = 0 \tag{4.5}
$$

$$
\sum_i \beta_i = 0 \tag{4.6}
$$

$$
\sum_i \phi_i = 0 \tag{4.7}
$$

Using the duality relation developed by Samuelson and by Shephard,[12] the relative share of factor i on the minimum expense locus is $\partial \ln C / \partial \ln P_i$. Thus, the factor-share equations from (4.3) are

$$M_K = \alpha_K + \gamma_{KK} \ln P_K + \gamma_{KL} \ln P_L + \gamma_{KE} \ln P_E + \gamma_{KN} \ln P_N + \beta_K \ln Q + \phi_K \ln T$$
$$(4.8)$$

$$M_L = \alpha_L + \gamma_{KL} \ln P_K + \gamma_{LL} \ln P_L + \gamma_{LE} \ln P_E + \gamma_{LN} \ln P_N + \beta_L \ln Q + \phi_L \ln T$$
$$(4.9)$$

$$M_E = \alpha_E + \gamma_{KE} \ln P_K + \gamma_{LE} \ln P_L + \gamma_{EE} \ln P_E + \gamma_{EN} \ln P_N + \beta_E \ln Q + \phi_E \ln T$$
$$(4.10)$$

$$M_N = \alpha_N + \gamma_{KN} \ln P_K + \gamma_{LN} \ln P_L + \gamma_{EN} \ln P_E + \gamma_{NN} \ln P_N + \beta_N \ln Q + \phi_N \ln T$$
$$(4.11)$$

For this general specification, the partial elasticities of substitution may be determined and the homothetic or nonhomothetic character of the underlying production function and the bias in technical change may be examined. Further, restrictions may be placed on the parameters in this specification to generate alternative specifications (i.e., impose homotheticity and/or neutral technical change).

The degree to which inputs may be substituted for one another is measured by the Allen partial elasticity of substitution.[13] The elasticity of substitution between inputs i and j is denoted as σ_{ij}. If σ_{ij} is less than or equal to zero, the inputs are complements. If σ_{ij} is positive, the inputs are substitutes, with a larger value of σ_{ij} indicating that the inputs are better substitutes for one another.

Uzawa[14] showed that the Allen partial elasticity of substitution between inputs i and j is $\sigma_{ij} = C[\partial^2 C / \partial P_i \partial P_j] / [(\partial C / \partial P_i)(\partial C / \partial P_j)]$. Using the translog derivatives, the cross elasticities of substitution are*

$$\sigma_{ij} = \frac{\gamma_{ij} + M_i M_j}{M_i M_j} \tag{4.12}$$

and the own elasticities are

$$\sigma_{ii} = \frac{\gamma_{ii} + M_i^2 - M_i}{M_i^2} \tag{4.13}$$

---

*The economic restriction of symmetry in the partial elasticities of substitution (i.e., $\sigma_{ij} = \sigma_{ji}$) requires that the parameters in the translog cost function be symmetric. That is, $\gamma_{ij} = \gamma_{ji}$.

Since the constant-output elasticity of demand for input i is $E_{ii} = M_i\sigma_{ii}$, each σ_{ii} must be nonpositive.

If the underlying production function is homothetic, it follows that changes in output should have no effect on relative input usage and therefore on factor shares (i.e., $\partial M_i/\partial Q = 0$); thus,

$$\beta_i = 0 \text{ for } i = K, L, E. \tag{4.14}$$

Therefore, homotheticity is rejected if any $\beta_i \neq 0$. A homothetic production function may then be specified by imposing the restriction $\beta_i = 0$ on equations (4.8) through (4.11).

Finally, if the production function is characterized by neutral technical change, changes in the level of technology should have no effect on relative input usage and factor shares (i.e., $\partial M_i/\partial T = 0$); and

$$\phi_i = 0 \text{ for } i = K, L, E. \tag{4.15}$$

If any $\phi_i \neq 0$, one may reject the hypothesis of neutral technical change. Technical change is factor using, neutral, or saving of the i-th input according as $\phi_i \gtreqless 0$. As in the case of homothetic production, an alternative specification with neutral technical change is obtained by imposing $\phi_i = 0$ on equations (4.8) through (4.11).

The major objectives of this analysis are summarized in equations (4.12) through (4.15). In the most general specification it is possible to distinguish between changes in input usage attributable to price-induced neoclassical substitution (equation [4.12]), nonhomothetic production (equation [4.14]), and biased technical change (equation [4.15]). Then, it is possible to examine the sensitivity of the estimates of the partial elasticities of substitution to alternative specifications by imposing the appropriate restrictions.

ESTIMATES FROM CANADIAN METAL MINING

In this section we shall examine empirically some of the properties of the production function for metal mining. Three questions are of primary interest:

1. Is the production function homothetic?
2. Has technological change been neutral or biased?
3. How much substitution among inputs is possible?

To these ends, four alternative specifications of the underlying production function are considered:

    I   Neither neutral technical change nor homotheticity imposed
   II   Neutral technical change imposed
 III   Homotheticity imposed
 IV   Both neutral technical change and homotheticity imposed

As shown in the preceding section, neutral technical change and homotheticity are imposed by restrictions on the parameters of $\ln Q$ and $\ln T$ in equations (4.8) through (4.11).

Since the factor shares in equations (4.8) through (4.11) sum to unity, the disturbances must sum to zero for each observation; consequently, the covariance matrix of the error terms is singular. While all parameters could be estimated by estimating any two of the three equations and incorporating restrictions (4.4) through (4.7), the parameter estimates may not be invariant to the equation deleted. Since maximum likelihood estimates would be invariant, the technique used is the iterative Zellner efficient (IZEF) method, which has been shown by Kmenta and Gilbert to yield maximum likelihood estimates.[15]

The parameters in equations (4.8) through (4.11) are overidentified in the sense that the estimates in each equation may not adhere to the restrictions imposed by symmetry and linear homogeneity of the cost function. These essential economic restrictions can be imposed using a method of staking the observations that is now well known.[16]

Equations (4.8) through (4.10) are estimated using the various specifications with symmetry and restriction (4.5) imposed. The remaining coefficients and their standard errors are calculated from restrictions (4.4), (4.6), and (4.7). The estimated parameters and their asymptotic standard errors are shown in Table 4.2. Examination of these results indicates clearly that the parameter estimates are sensitive to specification.

In the most general specification (i.e., neither homotheticity nor neutral technical change imposed), the estimates of $\hat{\beta}_i$ are consistent with the hypothesis that the underlying production function is homothetic. That is, it is not possible to reject statistically the null hypothesis that $\beta_i = 0$. The estimates of $\hat{\phi}_i$ indicate that technical change was capital using ($\hat{\phi}_K > 0$), labor saving ($\hat{\phi}_L < 0$), and neutral with respect to both energy and ore. Hence, these results would indicate that at least part of the change in the relative input usage was the result of biased technical change rather than neoclassical substitution.

If neutral technical change is imposed, homotheticity is clearly rejected. From these estimates it would appear that an expansion in output would use capital and ore ($\hat{\phi}_K, \hat{\phi}_N > 0$) and save labor and energy ($\hat{\phi}_L, \hat{\phi}_E < 0$).

If homotheticity is imposed, technical change appears to be not only capital-using and labor-saving but also ore-using. This result is entirely consistent with technical change that allows the extraction of minerals from lower-grade ore. It should be noted that, given the results from the most

TABLE 4.2

Estimated Coefficients, Four-Input Production

	I	II	III	IV
$\hat{\alpha}_K$	−2.25950* (0.56703)	−0.53494* (0.23432)	−1.98458* (0.43807)	−0.08460 (0.06905)
$\hat{\alpha}_L$	3.86630* (0.67892)	1.51485* (0.26347)	3.91961* (0.55851)	0.78482* (0.08282)
$\hat{\alpha}_E$	0.11277 (0.40165)	0.39274* (0.14613)	−0.17740 (0.32439)	0.02788 (0.03293)
$\hat{\alpha}_N$	−0.71957 (0.49506)	−0.37265 (0.19713)	−0.75763* (0.30332)	0.27190* (0.04219)
$\hat{\gamma}_{KL}$	−0.05722 (0.03330)	0.03771 (0.03177)	−0.04879 (0.03210)	0.11070* (0.01619)
$\hat{\gamma}_{KE}$	0.01612 (0.00931)	0.01975* (0.00821)	0.01744* (0.00838)	0.00143 (0.00803)
$\hat{\gamma}_{KN}$	−0.04754* (0.00720)	−0.04730* (0.00793)	−0.04777* (0.00704)	−0.05122* (0.00749)
$\hat{\gamma}_{LE}$	0.00387 (0.02709)	0.02642* (0.01262)	−0.01665 (0.02064)	0.00529 (0.00352)
$\hat{\gamma}_{LN}$	−0.07321* (0.00616)	−0.07375* (0.00945)	−0.07274* (0.00670)	−0.07427* (0.01249)
$\hat{\gamma}_{EN}$	−0.00806* (0.00119)	−0.00760* (0.00162)	−0.00839* (0.00192)	−0.00906* (0.00186)
$\hat{\gamma}_{KK}$	0.08864* (0.03753)	−0.01016 (0.03577)	0.07912* (0.03474)	−0.06091* (0.01585)

(continued)

TABLE 4.2 (Continued)

	I	II	III	IV
$\hat{\gamma}_{LL}$	0.12656* (0.04163)	0.00962 (0.03132)	0.13818* (0.03850)	-0.04172 (0.02284)
$\hat{\gamma}_{EE}$	-0.01193 (0.02516)	-0.03857* (0.01615)	0.00760 (0.01716)	0.00234 (0.00796)
$\hat{\gamma}_{NN}$	0.12881* (0.01871)	0.12865* (0.01807)	0.12890* (0.01855)	0.13456* (0.01976)
$\hat{\beta}_K$	-0.05297 (0.07803)	0.10254* (0.05222)	—	—
$\hat{\beta}_L$	0.05306 (0.08135)	-0.15625* (0.05219)	—	—
$\hat{\beta}_E$	-0.01341 (0.02193)	-0.04368* (0.02070)	—	—
$\hat{\beta}_N$	0.01332 (0.10904)	0.09739* (0.03025)	—	—
$\hat{\phi}_K$	0.76344* (0.26023)	—	0.60301* (0.13544)	—
$\hat{\phi}_L$	-0.99691* (0.26667)	—	-0.91698* (0.15882)	—
$\hat{\phi}_E$	0.01608 (0.09361)	—	0.06637 (0.08322)	—
$\hat{\phi}_N$	0.21739 (0.27575)	—	0.24760* (0.07328)	—

Asymptotic standard errors in parentheses.
*Significantly different from zero at the 95-percent confidence level.

TABLE 4.3

Estimated Elasticities, Four-Input Production

	I	II	III	IV
$\hat{\sigma}_{KL}$	0.65914*	1.22467*	0.70936*	1.65947*
	(0.19837)	(0.18925)	(0.19122)	(0.09647)
$\hat{\sigma}_{KE}$	2.07162*	2.31272*	2.15902*	1.09527*
	(0.61883)	(0.54548)	(0.55681)	(0.53377)
$\hat{\sigma}_{KN}$	−0.39498	−0.38778	−0.40162	−0.50288*
	(0.21121)	(0.23258)	(0.20645)	(0.21968)
$\hat{\sigma}_{LE}$	1.15518	2.05994*	0.33215	1.21220*
	(1.08694)	(0.50621)	(0.82793)	(0.14131)
$\hat{\sigma}_{LN}$	−0.29651*	−0.30607	−0.28834*	−0.31524
	(0.10904)	(0.16738)	(0.11869)	(0.22126)
$\hat{\sigma}_{EN}$	−0.59173	−0.50084	−0.65760	−0.79113
	(0.23549)	(0.31922)	(0.37923)	(0.36841)
$\hat{\sigma}_{KK}$	−1.26675*	−2.24188*	−1.36069*	−2.74276*
	(0.37039)	(0.35304)	(0.34285)	(0.15647)
$\hat{\sigma}_{LL}$	−0.44118*	−0.86168*	−0.39939*	−1.04628*
	(0.14971)	(0.11263)	(0.13844)	(0.08213)
$\hat{\sigma}_{EE}$	−25.50104*	−37.42593*	−16.75710*	−19.10983*
	(11.26354)	(7.23087)	(7.67989)	(3.56427)
$\hat{\sigma}_{NN}$	2.89569	2.88168	2.90429	3.39672
	(1.63174)	(1.57600)	(1.61745)	(1.72356)

Asymptotic standard errors in parentheses.
*Significantly different from zero at the 95-percent confidence level.

general specification, this specification appears to be the most appropriate for the metal-mining industry.

The estimated elasticities of substitution with their asymptotic standard errors are shown in Table 4.3. These values are evaluated at share sample means. The necessary condition for local stability of the derived demand function is satisfied for all inputs. That is, $\hat{\sigma}_{ii} \leq 0$ for i = K, L, E, N.*

---

*While the estimates of σ_{NN} are positive, they are not significantly different from zero (i.e., it is not possible to reject the hypothesis that $\sigma_{NN} = 0$). This estimation would imply that the constant

The effects of specification are more pronounced in these results. If neutral technical change is imposed, capital and labor become more substitutable and labor and energy appear to be substitutes rather than complements. The effects of the imposition of homotheticity are not as pronounced, but such a result is not surprising, given that homotheticity was not rejected in the most general specification. It is significant however that capital and energy appear to be much less substitutable if both restrictions are imposed.

Because of the difficulties encountered in obtaining data for ore, both in the sense of a homogenous input over time and in that of a relevant market price for the ore, we felt that it might be advisable to consider a three-input setting to determine whether the inclusion of the ore input affected the estimates. We should note that the preceding results do not preclude this application of weak functional separability. As Berndt and Christensen showed,[17] weak separability is indicated if the Allen partial elasticities of substitution between a factor in the separable set and some factor outside the set are equal (i.e., $\sigma_{KN} = \sigma_{LN} = \sigma_{EN}$). Since Blackorby, Primont, and Russell and Denny and Fuss indicated that such separability is consistent with homotheticity in the translog functional form, it is in specification IV that this test is most appropriate.[18] Since Table 4.3 indicates that $\hat{\sigma}_{KN}$, $\hat{\sigma}_{LN}$, and $\hat{\sigma}_{EN}$ cannot be shown to be significantly different, weak separability of ore from the other inputs is not rejected.

The parameter estimates from the relative factor share equations in the three-input setting, and their asymptotic standard errors are presented in Table 4.4. Not only are the conclusions concerning homotheticity and the bias in technical change the same as in the four-input setting; the magnitudes of the parameter estimates are also very similar.

The estimated partial elasticities of substitution derived in this three-input model and their asymptotic standard errors are shown in Table 4.5. Again, as a result of the imposition of neutral technical change, capital and labor become more substitutable, and labor and energy are substitutes rather than complements. The most interesting feature, however, is the fact that, if both neutral technical change and homotheticity are imposed, capital and energy appear to be complements rather than substitutes.

---

output-price elasticity of the demand for ore (in this case the tax elasticity) is zero. That is, for a given level of output of ore concentrate, an increase in the user cost of ore would have no effect on the amount of ore extracted. Such a result is not implausible, certainly in the context of the weak separability, which will be demonstrated. However this evidence would also indicate that, over the period considered, there has not been significant high-grading of ore in response to changes in the mineral tax.

TABLE 4.4

Estimated Coefficients, Three-Input Production

	I	II	III	IV
$\hat{\alpha}_K$	−2.89066* (0.55767)	−0.84039* (0.24000)	−2.56047* (0.43359)	−0.05971 (0.07423)
$\hat{\alpha}_L$	3.57026* (0.64498)	1.32471* (0.23537)	3.62719* (0.58224)	1.04224* (0.06863)
$\hat{\alpha}_E$	0.32040 (0.41352)	0.51568* (0.15875)	−0.06672 (0.37014)	0.01747 (0.04313)
$\hat{\gamma}_{KL}$	−0.09179* (0.03507)	0.01255 (0.03479)	−0.08020* (0.03421)	0.10725* (0.01612)
$\hat{\gamma}_{KE}$	0.01278 (0.01049)	0.02206* (0.01000)	0.01302 (0.00949)	−0.00877 (0.01049)
$\hat{\gamma}_{LE}$	0.01483 (0.02846)	0.03233* (0.01414)	−0.01353 (0.02366)	−0.00318 (0.00316)
$\hat{\gamma}_{KK}$	0.07901 (0.04099)	−0.03461 (0.04037)	0.06718 (0.03768)	−0.09848* (0.01761)
$\hat{\gamma}_{LL}$	0.07696 (0.04266)	−0.04488 (0.03178)	0.09373* (0.04159)	−0.10407* (0.01817)
$\hat{\gamma}_{EE}$	−0.02761 (0.02569)	−0.05439* (0.01732)	0.00051 (0.01871)	0.01195 (0.01000)
$\hat{\beta}_K$	−0.05566 (0.07078)	0.15941* (0.05577)	—	—
$\hat{\beta}_L$	0.07758 (0.06181)	−0.10241* (0.05089)	—	—
$\hat{\beta}_E$	−0.02192 (0.02345)	−0.05700* (0.02280)	—	—
$\hat{\phi}_K$	0.95188* (0.25495)	—	0.77162* (0.13820)	—
$\hat{\phi}_L$	−0.93444* (0.23537)	—	−0.81368* (0.16673)	—
$\hat{\phi}_E$	−0.01744 (0.10000)	—	0.04206 (0.09539)	—

Asymptotic standard errors in parentheses.
*Significantly different from zero at the 95-percent confidence level.

Source:

79

TABLE 4.5

Estimated Elasticities, Three-Input Production

	I	II	III	IV
$\hat{\sigma}_{KL}$	0.56471* (0.16631)	1.05951* (0.16496)	0.61968* (0.16221)	1.50860* (0.07647)
$\hat{\sigma}_{KE}$	1.67597* (0.55475)	2.16682* (0.52893)	1.68867* (0.50179)	0.53613 (0.55474)
$\hat{\sigma}_{LE}$	1.47621 (0.91390)	2.03815* (0.45412)	0.56554 (0.75989)	0.89789* (0.10154)
$\hat{\sigma}_{KK}$	−1.17769* (0.32017)	−2.06520* (0.31536)	−1.27010* (0.29435)	−2.56411* (0.13753)
$\hat{\sigma}_{LL}$	−0.47519* (0.12282)	−0.82596* (0.09150)	−0.42691* (0.11975)	−0.99637* (0.05230)
$\hat{\sigma}_{EE}$	−27.81735* (9.20125)	−37.40524* (6.20344)	−17.74242* (6.70050)	−13.64509* (3.58156)

Asymptotic standard errors in parentheses.
*Significantly different from zero at the 95-percent confidence level.

Our results can be summarized as follows. It appears that the primary substitution occurred between capital and energy, with capital and labor being somewhat less substitutable; absolute complementarity exists between the other input pairs. The estimated parameters of the underlying cost function were consistent with a homothetic production function. With homotheticity imposed, the bias in technical progress was found to be toward using capital and ore and saving labor; with respect to energy, technological change was neutral. In the period under consideration the price of labor rose rapidly and the price of capital rose moderately relative to the prices of energy and ore. The data for input usage indicate that the industry substantially reduced its usage of labor relative to the other inputs and moderately reduced its usage of capital relative to energy. If biased technical change is not considered, such behavior might be interpreted as evidence of neoclassical substitution. However, the results of this analysis would tend to indicate that the reduction in labor usage resulted more from biased technical change than from substitution. The modest reduction in the usage of capital relative to energy would, however, be attributable to the strong substitutability between these inputs.

A NOTE ON NONMETAL MINING

In the Canadian nonmetal mining industry data were available only for capital, labor, and electric-energy inputs. The trends in relative factor usage and in factor prices were very similar to those for metal mining during the period 1962–74.[19]

Using the same methodology as that employed for metal mining, the three-input, translog factor-share equations were estimated. Only the most general form was estimated (i.e., imposing neither neutral technical change nor homotheticity). The results of this estimation are presented in Table 4.6. As in the case of metal mining, the results are consistent with an underlying production relation that is homothetic (i.e., $\hat{\beta}_i$ is not significantly different from zero for i = K, L, E). However, in contrast to the results for metal mining, technical change appeared to be not only capital using and labor saving but also energy saving.

The estimated elasticities of substitution and their asymptotic standard errors are presented in Table 4.7. While the conditions for local stability are satisfied (i.e., $\sigma_{ii} \leq 0$ for i = K, L, E), the estimated positive but insignificant parameters for σ_{KK} and σ_{LL} require that these results be discounted somewhat.* Indeed, in marked contrast to our results for metal mining, the substitutability is only between labor and energy rather than between capital and labor and capital and energy. This caveat notwithstanding, these results might imply that changes in the relative levels of input usage in nonmetal mining have been brought about through induced technical change rather than through neoclassical substitution.

SUMMARY AND IMPLICATIONS

In this empirical analysis we have employed a generalized cost function to examine some of the properties of the underlying production relations for mineral extraction. Using data drawn from metal mining in Canada for the period 1962–74, we found several important features. Changes in the scale of operations would not be expected to affect the relative usages of inputs. In other words changes in the level of production would not change the ratios of input usage, even though the levels of usage would change.

---

*These results would imply that the constant output-demand functions for capital and labor are price inelastic. Such a proposition is clearly less plausible than the preceding finding that the constant output demand function for ore is price (tax) inelastic.

TABLE 4.6

Estimated Coefficients, Three-Input Production Canadian Nonmetal Mining

$\hat{\alpha}_K$	−6.24833* (1.74642)	$\hat{\gamma}_{LE}$	0.02386 (0.01414)	$\hat{\beta}_L$	0.07203 (0.08888)
$\hat{\alpha}_L$	6.49704* (1.61245)	$\hat{\gamma}_{KK}$	0.31490* (0.08820)	$\hat{\beta}_E$	0.02757 (0.01549)
$\hat{\alpha}_E$	0.75129* (0.30000)	$\hat{\gamma}_{LL}$	0.24812* (0.07457)	$\hat{\phi}_K$	2.06806* (0.62450)
$\hat{\gamma}_{KL}$	−0.27198* (0.08025)	$\hat{\gamma}_{EE}$	0.01906 (0.01000)	$\hat{\phi}_L$	−1.81894* (0.57359)
$\hat{\gamma}_{KE}$	−0.04292* (0.01414)	$\hat{\beta}_K$	−0.09960 (0.09757)	$\hat{\phi}_E$	−0.24912* (0.10488)

Asymptotic standard errors in parentheses.
*Significantly different from zero at a 95-percent confidence interval.

TABLE 4.7

Estimated Elasticities, Three-Input Production Canadian Nonmetal Mining

$\hat{\sigma}_{KL}$	−0.19096 (0.35140)
$\hat{\sigma}_{KE}$	−1.01612 (0.66431)
$\hat{\sigma}_{LE}$	2.31927* (0.78195)
$\hat{\sigma}_{KK}$	0.24270 (0.32813)
$\hat{\sigma}_{LL}$	0.00858 (0.38433)
$\hat{\sigma}_{EE}$	−12.04922* (5.93118)

Asymptotic standard errors in parentheses.
*Significantly different from zero at a 95-percent confidence interval.

Our results also show that technological change over this period appears to have been biased. The estimates indicated that technological change resulted in a relative increase in the usage of capital and ore and a relative decrease in the usage of labor. We did not distinguish between technological change resulting from innovation and that from switches to alternative types of available technology.

Finally, we tested for the existence of substitutability between inputs used in mineral extraction. The results indicate that capital and energy were most substitutable for one another. This certainly has important implications for the future. We also deduced that capital and labor were substitutes but far less so that capital and energy. There was no substitutability between any other pair of inputs.

Our results have substantial implications both for considering possible future trends and for analyzing the potential effects of alternative governmental policies. For instance, if the current trend of rising energy prices relative to the prices of the other inputs continues, our estimations would indicate a further increase in the usage of capital relative to that of energy. Moreover, if the trend existing during the period prior to 1973 (when the price of labor increased relative to that of the other inputs) is reestablished, we might expect to see additional capital deepening, both from substitution and from induced technical change. To the extent that there continue to be changes in technology, our results would indicate an increase in the relative use of capital and ore and a decrease in the relative use of labor.

In following chapters we will consider the effects of changes in governmental tax and environmental legislation. In general, we assume that these are levied on the firm in such a way as to leave relative prices of inputs unchanged. However, this need not necessarily be the case. Clearly, governmental policies that would have the effect of changing the relative price of energy or labor (e.g., taxes on or rationing of energy and minimum-wage legislation) or altering the rate of technical growth (e.g., changing the tax treatment of research expenditures) would result in changes in relative input use as outlined in the preceding paragraphs. If, however governmental policies increase the price of capital services (e.g., a reduction in the average lifetime of capital for tax purposes), the results of this analysis would indicate that, with the relative decrease in capital usage, the relative usage of labor and energy would be expected to increase. These are quite important policy implications.

NOTES

1. Ernst R. Berndt and David O. Wood, "Technology, Prices, and the Derived Demand for Energy," *Review of Economics and Statistics* (August 1975): 259–68. Specifically, they used a transcendental logarithmic cost function to obtain their estimates. They did, however, impose homotheticity and neutral technical change on the underlying production function.

2. E.R. Berndt and D.W. Jorgenson, "Production Structure," in D.W. Jorgenson, E.R. Berndt, L.R. Christensen, and E.A. Hudson, *U.S. Energy Resources and Economic Growth*, Final Report to the Ford Fountain Energy Policy Project, Washington, D.C., October 1973.

3. M.A. Fuss, 'The Demand for Energy in Canadian Manufacturing: An Example of the Estimation of Production Structures with Many Inputs," *Journal of Econometrics* (January 1977): 89–116.

4. M. Denny, J.D. May, and C. Pinto, "The Demand for Energy in Canadian Manufacturing: Prologue to an Energy Policy," *Canadian Journal of Economics* (May 1978): 300–13. They obtained their estimates using a generalized Leontief cost function rather than the transcendental logarithmic cost function employed by the other researchers and to be employed in our analysis.

5. James M. Griffin and Paul R. Gregory, "An Intercountry Translog Model of Energy Substitution Responses," *American Economic Review* (December 1976): 845–57.

6. Griffin and Gregory felt that their results were indicative of long-run response while those of Berndt and Wood and others tended to capture short-run relations. Berndt and Wood (Ernst R. Berndt and David O. Wood, "Engineering and Econometric Approaches to Industrial Energy Conservation and Capital Formation: A Reconciliation," working paper, University of British Columbia, December 1977) point out that the use of too limited a subset of the inputs may bias the estimates; thus, the results of Griffin and Gregory may be the result of the exclusion of intermediate materials (a point they acknowledged).

7. Laurits R. Christensen and William H. Greene, "Economies of Scale in U.S. Electric Power Generation," *Journal of Political Economy* (August 1976): 665–76.

8. For a listing of the data and a more complete description of the manner in which the variables were calculated, the reader is referred to the technical report prepared by Smithson et al., *op. cit.*

9. D.W. Jorgenson and Z. Griliches, "The Explanation of Productivity Change," *The Review of Economics and Statistics* (July 1967): 249–84.

10. E.G. Furubotn, "The Orthodox Production Function and the Adaptability of Capital," *Western Economic Journal* (Summer 1965): 288–300; Furubotn, "Long-Run Analysis and the Form of the Production Function," *Economia Internazionale* (February 1970): 3–35.

11. L.R. Christensen, D. Jorgenson, and L. Lau, "Conjugate Duality and the Transcendental Logarithmic Production Function," presented at the Second World Congress of the Econometric Society, Cambridge, England, 1970.

12. P.A. Samuelson, *Foundations of Economic Analysis* (Cambridge, Mass.: Harvard University Press, 1947); R.W. Shephard, *Theory of Cost and Production Functions* (Princeton: Princeton University Press, 1970).

13. For the derivation of this parameter, see R.G.D. Allen, *Mathematical Analysis for Economists* (London: Macmillan, 1938).

14. H. Uzawa, "Production Functions with Constant Elasticities of Substitution," *Review of Economic Studies* (October 1962): 291–99.

15. J. Kmenta and R. Gilbert, "Small Sample Properties of Alternative Estimators of Seemingly Unrelated Regressions," *Journal of the American Statistical Association* (December 1968): 1180–1200.

16. See D.B. Humphrey and J.R. Moroney, "Substitution among Capital, Labor, and Natural Resource Products in American Manufacturing," *Journal of Political Economy* (February 1975): 57–82; Christensen and Greene, *op. cit.*

17. E.R. Berndt and L.R. Christensen, "The Internal Structure of Functional Relationships: Separability, Substitution, and Aggregation," *Review of Economic Studies* (July 1973): 403–10.

18. C. Blackorby, D. Primont, and R.R. Russell, "On Testing Separability Restrictions with Flexible Functional Forms," *Journal of Econometrics* (March 1977): 195–209; M. Denny and M. Fuss, "The Use of Approximation Analysis to Test for Separability and the Existence of Consistent Aggregates," *American Economic Review* (June 1977): 404–18.

19. For a description of this data, the reader is referred to the technical report prepared by the authors: Smithson et al., *op. cit.*

5

Assimilation of
New Technology

Until now we have generally assumed a constant technology in mineral extraction, with the exception of that part of Chapter 4 in which we estimated the effects of technological change on relative input usage. In this chapter we shall consider in depth how to examine the economic feasibility of a new technology or a technological change in mineral extraction. Because of the importance of this aspect of mineral economics we devote an entire chapter to the subject. While this chapter is fairly technical in its economic analysis, only the section in which the mathematical model is derived will be particularly difficult for the reader who is not an economist or is not mathematically trained. For those readers the results are summarized at the conclusion of that section. The remainder of the chapter relies primarily on graphical analysis.

Because of the increased interest in the subject, we shall concentrate our analysis primarily on the use of new, nonconventional techniques in the extraction of resources. One example, which is receiving attention is deep-seabed mining—an alternative technology—to add to the supply of minerals forthcoming from extraction that uses conventional techniques. Other

This chapter is based on a report prepared by two of the authors, W.P. Gramm and S.C. Maurice, "The Economic Feasibility of Peacetime Nuclear Explosives in Mineral Extraction," in *PNE* (*Peaceful Nuclear Explosives*) *Activity Projections for Arms Control Planning*, a report submitted to the U.S. Arms Control and Disarmament Agency by Gulf Universities Research Consortium, January 20, 1975.

examples are the use of nonconventional techniques in extracting gas from gas fields of such low permeability that they could not be exploited with techniques presently used and of recovering oil from oil shale. It is possible that these deposits could be exploited using nonconventional techniques, such as massive hydrolic fracturing or stimulation with nuclear devices in the case of gas or a modified garrett process or nuclear *in situ* retorting in the case of oil shale. Similar conditions are said to exist in copper leaching.

In any case, most previous research has been involved with consideration of merely technological rather than economic feasibility. This is a major gap in the analysis. While it might be technologically possible to employ some nonconventional extraction technique, it is possible that the price of the resource to be extracted is too low or the price of the nonconventional technology is too high for the technique to be economical. In this study we will concentrate on the economic conditions under which a profit-maximizing firm would be induced to employ some nonconventional extraction technique.

The analysis will consider the economic feasibility of nonconventional extraction techniques in general rather than that of some specific technique. Thus, our conclusions can be applied to any specific technological innovation of interest to the reader. We begin with a brief discussion of some previous approaches to the determination of the feasibility of nonconventional techniques. As will be shown, much of this work failed to incorporate some fundamental economic principles. Then we examine the issue using simple graphical techniques to show both the effect of the market price on the demand for and supply of the resource and the effect of different prices for the nonconventional technology on supply. This section shows how to examine the effect on the natural resource price of the use of nonconventional extraction techniques. In the following section we develop a more rigorous mathematical model to examine the behavior of a profit-maximizing firm. This model is an extension of the basic model developed in Chapter 2. We will use it to determine the firm's equilibrium level of usage of the nonconventional technology and to examine the effects of changes in some economic parameters on the level of usage. Finally, we provide an outline of the methodology in which our model could be used to evaluate the economic feasibility of a particular type of nonconventional technology. As always we shall indicate the parts of the chapter that can be skipped without loss of continuity by the reader who is not interested in the rigorous economic or mathematical techniques. We shall always summarize the results in non-technical language.

AN EVALUATION OF PREVIOUS APPROACHES

The economic analysis of the feasibility of technological innovations in mineral extraction has been sketchy at best, misleading and incorrect at worst. In this section we will omit the technological considerations and will instead concentrate upon a brief summary of the economic implications that have been set forth.

Most economic, as opposed to technological, analyses of the feasibility of technological innovation take a rather standard approach. They generally assume that for some reason or other the quantity supplied of a resource will not meet the demand for the resource during some future period. Therefore, according to the analysts, there will be a shortage of the resource in the future, and some individuals or firms will simply have to do without unless additional supplies are found. Having established a future shortage, these analyses go on to point out that additional large amounts of the resource are available, but technological conditions are such that these areas are not economically exploitable with existing techniques at predicted prices.

Clearly, there is a great deal of uncertainty involved in estimating the impact of these new techniques. But the uncertainty is usually assumed to be most prevalent in the case of technological costs. Demand, on the other hand, is generally assumed to be known. The amount of the resource that will be "needed" during some future period is predicted, usually with a great deal of confidence. Or it is stated that "self sufficiency" (again a term that is ambiguous at best) will require a certain amount of the resource. Next, the amount "needed" is compared with the amount that will be available (again ambiguous), and the amount of the shortage is estimated.

Then those attacking the problem attempt to determine whether nonconventional methods of production can "fill the gap" at a cost low enough to compete with conventionally produced resources. Conditions under which this would be the case are determined. Usually, it is assumed that the price consumers pay for a resource is solely determined by the cost of recovering, transporting, and distributing it. The "gapmanship" technique frequently assumes that the higher prices of the portions of the natural resource that are produced nonconventionally are simply passed along to the consumer. Those using this approach then estimate the effect the higher-priced resource will have upon the average price that consumers will have to pay.

An example should clarify this point. Suppose someone says that more gas will be needed on the West Coast in the future. The amount of the shortage is estimated, and it is assumed that this shortage can be alleviated by gas produced in the Rocky Mountains with nuclear stimulation. This supplementary gas has a higher production cost than conventionally produced gas. However, gas produced from both sources is shipped in the same pipelines. Assume that 10 percent of the gas shipped is nuclearly stimulated gas, extracted at a higher cost. This higher-cost gas raises the weighted average cost of the gas in the pipeline. That is, if the average cost of a unit of conventionally produced gas is C_C, and the average cost of nuclearly stimulated gas is C_N, the average cost rises to $0.9C_C + 0.1C_N$. The amount of increase depends, of course, upon the difference in cost and the ratio of conventionally produced to nuclearly stimulated gas in the pipe. In any case, the increased cost is simply assumed to be passed along to consumers in the form of a price increase.

Almost every economic analysis of the possibility of using new technologies to extract resources has been an *ad hoc* evaluation of a specific project or group of projects. There has been no attempt to take a general approach to the problem based upon universally (or a least generally) accepted economic principles. Even in the more specifically oriented economic studies there is usually little or no fundamental economic methodology.

To be more specific, there are four fundamental errors in logic in most of these types of analyses. First, it is usually implicitly assumed that price has no effect upon the consumption of output. Second, the analysts fail to consider that, in a freely functioning market, after some period of adaptation there will be no shortage (or gap) if government permits price to vary. Third, they ignore the fact that the amount of a resource required for self-sufficiency depends upon the price. Fourth, they fail to incorporate the fact that the quantity that private producers will supply to the market depends upon, among other things, the price of the output. Clearly, these points, described in detail in Chapter 2, must be, but have not been, considered in any economic analysis of the possibility of using new technologies.

As we noted in Chapter 2, there does exist a fixed amount of any resource in the ground, and no more will be added. But, as price rises, suppliers would find it profitable to extract reserves that would be unprofitable at lower prices. With higher resource prices higher cost technologies will become profitable, and additional exploration will be warranted. While we may be reasonably certain that quantity demanded varies inversely with price, and quantity supplied varies directly with price, the degree of responsiveness depends respectively upon the elasticities of demand and supply. This is truly the point at issue, yet there has been little attention paid. In our analysis we will concentrate upon the basic question of the responsiveness of supply and demand.*

---

*The primary determinants of the responsiveness (i.e., price elasticities) are discussed in Chapter 2, pp. 24–29.

A SIMPLE GRAPHICAL ANALYSIS

Let us begin with a simplified version of the industry demand and supply functions described in Chapter 2. In Figure 5.1 the demand and supply functions are labeled DD and SS, respectively, where the supply function reflects conventional production only. The resulting market equilibrium would occur at P, Q.

There exist fields or deposits that could not be profitably exploited using conventional techniques under the expected set of relative prices. For example, a gas field may be of such low permeability that the costs of extraction would not be covered by expected revenue at any level of output. These fields would not even be specified as existing reserves. But it may be possible to lower the cost of extraction in these areas by using nonconventional methods, such as nuclear stimulation or some other technique.

A hypothetical situation is depicted in Figure 5.2. Let the expected price in a period be such that total revenue (price times quantity) is given by the ray from the origin labeled "Total Revenue" (TR). Quantity per period is measured along the horizontal axis and revenue in dollars, along the vertical. The slope of the ray is the price. Let total cost, using conventional methods of extraction, of producing each level of output be given by the curve FC_1. If the cost curve becomes practically vertical at some level of output, this rate of output sets the limit to the field under conventional technology. Fixed cost (or front-end cost) is OF; the remainder is variable or operating cost. The field would not be exploited under any circumstances under these conditions, because revenue would not cover costs at any level of output. Of course, at some higher price the total revenue curve may rise sufficiently for the field to be profitable at some levels of output.

Suppose, however, that hydrolic fracturing could lower the total cost curve to HC_H. Since HC_H is everywhere greater than total revenue with the given price, this method of extraction would not be chosen. Finally, the use of nuclear stimulation may drop the cost function still lower to NC_N. At levels of extraction between Q_1 and Q_2 total revenue exceeds total cost, and some profits would be attainable. The firm would choose that level of output at which the distance between revenue and cost was greatest.*

Therefore, in the case shown in Figure 5.2 nuclear devices would be used, since this method is the only profitable one under the given set of

---

*As shown in Chapter 2, this requires producing that output at which marginal cost equals marginal revenue. We might note that the most profitable level of output need not be the level that is technologically optimal—that is, the rate of extraction that gives maximum total yield.

FIGURE 5.1

Market Equilibrium Using Only Conventional Production

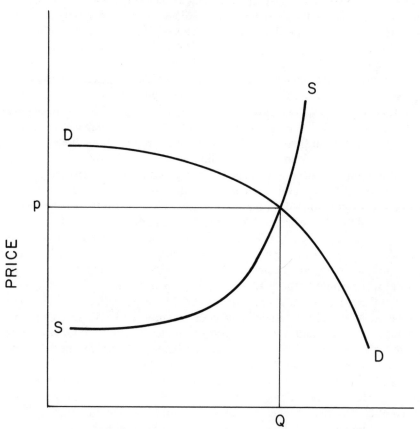

QUANTITY PER PERIOD OF TIME

FIGURE 5.2

Output Using Alternative Production Technologies

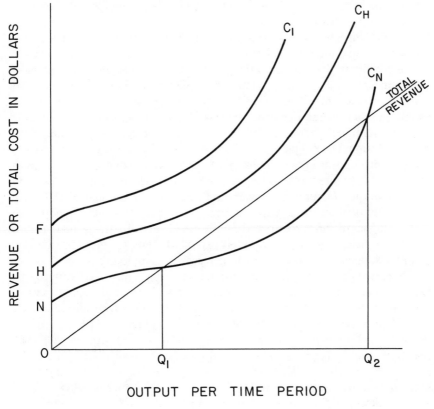

OUTPUT PER TIME PERIOD

circumstances. Fixed cost is ON, while variable cost is the difference between NC_N and fixed cost. We might emphasize in passing that while cost may be lowest when nonconventional technology is used in the field in question, in other fields with different formation characteristics cost could be far lower if conventional methods were employed. Nonconventional methods would only be economically justified when conventional technology is not economically feasible.

To summarize, in making decisions or predictions about the future, policy makers would wish to know the answers to three general questions. First, they would wish to know whether any nonconventional method will be economically feasible under any relevant expected set of circumstances. These expected circumstances are the projected prices of inputs, including the price of the innovation, and the price of the output. Second, they would wish to compare the costs of all nonconventional methods of extraction to see which, if any, will be chosen. Third, they would want to determine the future price and output if nonconventional methods of extraction are added to conventional methods.

Again, let us first approach the problem theoretically, using the simplest graphical techniques. Assume at first for simplicity that only one alternative to conventional techniques is feasible at any relevant set of relative prices. A certain number of resource deposits or fields are known to be noneconomic with conventional methods, but they may be profitably exploited with nonconventional techniques at a certain level of prices during a given period. These given deposits or fields are assumed to have cost configurations such as those given in Figure 5.2. At a given resource price the firm will produce a given level of output (assuming profitability) and use an optimal level of the nonconventional technology to produce this output. We would hypothesize that the higher the resource price, the more each firm will extract, and the more deposits or fields that can be profitably exploited. Thus, for the entire market, quantity supplied will increase with increases in commodity price.

Since the price of the nonconventional technique is included in total costs, higher prices for this technology shift nonconventional market supply upward. Thus, there is a market supply curve of nonconventionally produced output for each price of the nonconventional technology. In Figure 5.3 three out of the family of such supply curves are shown. $S_3 S_3$ is associated with a higher price than $S_2 S_2$, and $S_2 S_2$ with a higher price than $S_1 S_1$. Note that if the price of the nonconventional technology is such that $S_1 S_1$ is the relevant supply, no nonconventional production will occur if commodity price is below OS_1. If $S_2 S_2$ is relevant, OS_2 is the relevant cut-off price, and so forth. Therefore, the expected price of the technology and the expected resource price determine whether any output will be produced using nonconventional techniques.

Now we can combine the supply curves for conventionally produced output with the supply of nonconventionally produced output. The total market supply is the horizontal summation of both supply curves. At a given price for the nonconventional technology the conventional supply is the only effective supply until a resource price is attained at which it becomes profitable to use nonconventional technology. At higher resource prices the total amount supplied at each price is the amount supplied conventionally plus the amount supplied nonconventionally. In Figure 5.4 let SS be the supply curve considering only conventional methods. At the lowest price for nonconventional technology the segment AB is added to SS and the market supply is SAB. At a higher price the segment CD is added and supply becomes SCD; at a still higher price of the alternate technology supply is SEF. The higher the price of

FIGURE 5.3

Supply Curves for Alternative Prices of the Nonconventional
Technology

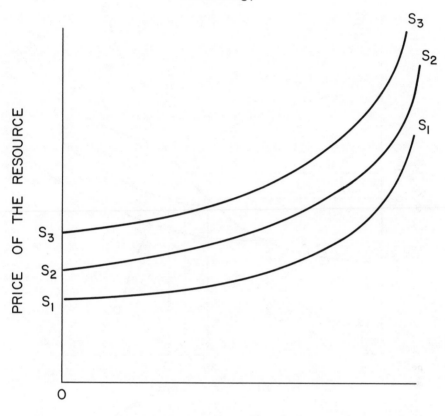

OUTPUT PER PERIOD OF TIME
PRODUCED WITH NON - CONVENTIONAL
TECHNIQUES

FIGURE 5.4

Market Equilibrium Using Conventional and Nonconventional
Production

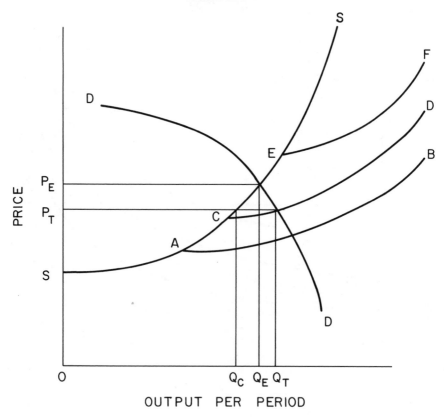

the innovation, the greater the resource price must be for this method of
production to be utilized at all.

With no nonconventional production and with free markets, the
intersection of demand, DD, and conventional supply, SS, determines an
equilibrium price of P_E and output of Q_E. If the price of the given alternative
technology is such that the segment CD becomes economically feasible, supply
is SCD. If this method is used, output price falls to P_T and output rises to Q_T.

Of this output, Q_C is produced conventionally, and the remainder $(Q_T - Q_C)$, shown as the distance $Q_C Q_T$, is produced with nonconventional techniques. We note that in this case, while total output increases, the fall in price necessary to sell the increased production actually decreases conventional production. *In other words, use of nonconventional technology cannot increase the price of the resource; it will only decrease that price.* There are, of course, cases in which nonconventional techniques would increase production and decrease resource price without decreasing conventional production, although this circumstance would probably be rather unusual. Nonetheless, Figure 5.5 shows such a case. The demand curve, DD, and conventional supply SS, determine price and output in equilibrium where they intersect. Note that SS is perfectly inelastic (vertical) over the relevant range. That is, Q_C appears to be a finite limit for conventionally produced output, at least in the short run. Let the price of the nonconventional technology be such that the portion CD is added to supply. Price falls; total output rises to Q_T, but conventional production remains constant at Q_C, the difference, $Q_C Q_T$, being made up entirely of nonconventional production.

Let us return to Figure 5.4. If the price of the alternate technology falls so that SAB becomes relevant, total output increases and price falls. If one could predict demand and conventional supply in some future period, he could predict—perhaps approximately—the price and production in that period assuming no nonconventional production. Next, one would estimate the highest price of the alternative technology that would cause nonconventional supply to fall below this predicted price over some range of output. This price would be the maximum price at which this nonconventional technique would be used. Total supply would be estimated for several prices of the innovation below this maximum price and within the technologically feasible range. For each augmented supply, there would be a predicted resource price and predicted total output. After one estimated the part of the total that would be produced nonconventionally, technological characteristics would determine the usage level of nonconventional techniques associated with this level of production. Thus, one could gauge a level of use of the alternate technology associated with each price of the innovation and in this way derive a demand for the technological innovation.

Now clearly, the data are not going to be such that so precise a method of analysis can be used. The preceding analysis simply shows a beginning framework of analysis and sets forth the problem in a framework other than the "gapsman" context. To stress one point developed in this analysis, *under a free market, nonconventional production will lower resource prices rather than raise them, contrary to what had been assumed in most previous economic analyses.* Costs do have an effect upon prices, but demand has an effect also. In the long run, in order to sell more, price must generally fall, unless an externally

FIGURE 5.5

The Requirement for Conventional Production to be Unchanged
after Nonconventional Technology is Implemented

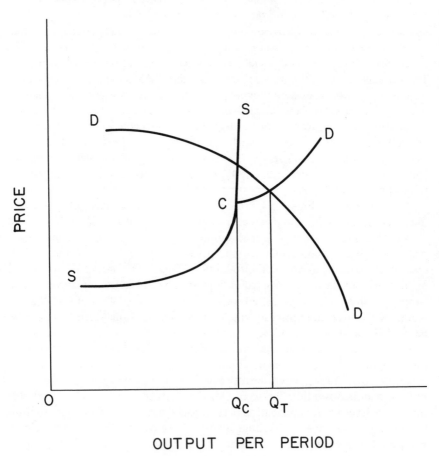

OUTPUT PER PERIOD

caused shortage (possibly the result of a governmental ceiling price below equilibrium) previously existed.*

A MATHEMATICAL APPROACH

In this section we will expand the mathematical model developed in Chapter 2 to consider the demand for technological innovations over some future time period and the effects upon output and price. In the following subsections we develop the model itself, examine the equilibrium conditions, and, finally, examine the theoretical implications of the model and its comparative statics properties. The reader who is not mathematically inclined or not interested in the derivations may easily omit the mathematical analysis. The results that are important for our analysis are summarized verbally in the final subsection, "A Brief Overview of the Mathematical Analysis."

The Basic Model

For any field or deposit of a resource the production function is such that there are several techniques possible for the exploitation of that field. Those areas for which conventional technologies would be economically feasible— i.e., profitable—are counted as reserves. No known technique would be more efficient in these areas than conventional methods. There are some fields or deposits that have large quantities of a resource, but the formation characteristics of these areas is such that they could not be economically exploited with conventional technologies. Our task is to determine the conditions, if any, under which it would be economically feasible to extract the resources with known nonconventional techniques. We wish, here, to develop a general model on which to base analysis.

Suppose that a firm has access to one of these fields or deposits. Assume for simplicity that there is only one nonconventional technique available. The firm has two decisions to make. First, it must decide whether to produce at all

---

*In a situation where the government created a shortage by imposing a ceiling price on the resource, it might be that the government would want to eliminate the shortage through nonconventional extraction. If the price of the nonconventional technology is so high that the nonconventionally produced supply lies everywhere above the ceiling price, then for production to be increased through nonconventional extraction, it is necessary that the ceiling price be raised. Hence, only in the case of an artificially low resource price would nonconventional extraction lead to higher resource prices.

with nonconventional techniques. Second, if it decides to produce any output, the firm must decide at what rate to produce over its time horizon.

In order to incorporate the effects of nonconventional technology, the mathematical model developed in Chapter 2 must be expanded to include the effects of nonconventional technology on operating costs, the direct expenditures on the nonconventional technology, any additional costs that might accrue if the nonconventional techniques were hazardous (this cost is very relevant in the case of nuclear stimulation), and taxes. Including these features, the present value of the firm is

$$PV = \sum_{t=0}^{H} \left(\frac{1}{1+r}\right)^t [R_t - C_t(q_t, q_{t-1}, \ldots, q_0, N_t, N_{t-1}, \ldots, N_0, \mu)$$

$$- F_t - P_{N_t} N_t - G(N_t) - T_t] \tag{5.1}$$

where H is the time horizon over which the firm makes its plans.

r is the rate that is used by the firm to discount future profits.

R_t is revenue after royalty in period t. It may be defined as $R_t = (1 - L)P_t q_t$

where P_t is the expected price of the resource in period t,

q_t is the output (in barrels, cubic feet, or tons per period) of the resource in period t, and

L is the proportion of total revenue that is applied toward land costs.

C_t denotes operating costs in period t. C_t includes both fixed and variable elements of operating costs. It is a function of both output in period t and output in all periods prior to t. It is also a function of nonconventional technology employed in period t and in all periods prior to t and of geographic features of the deposit.

N_t is the amount of nonconventional technology used in the t-th period.

μ is the formation characteristic of the field or deposit. Depending on the resource being considered, it is a function of, among other things, depth, richness, water, and permeability. This is a parametrically given variable.

F_t is the capital cost (i.e., fixed cost) in the t-th period. These capital costs will be both front-end costs and capital additions. We assume that capital costs are parametrically given to the firm. These costs will occur

G(N$_t$)   even if regulatory time delays occur. It is, therefore expenditure on capital additions, less depreciation and depletion.

G(N$_t$)   is the hazard cost. This cost is a function of the level of nonconventional technology employed.

P$_{N_t}$ is the cost of the nonconventional technology in period t. We separate this cost from other costs because we will want to isolate specifically the effect of changes in the price of the alternate technology. Certainly we do not know how the IRS would treat this cost for taxation purposes; it might consider this a capital expenditure. For now, however, we treat this as a general cost.

T$_t$ is federal income tax in the t-th period defined to be

$$T_t = A[R_t - C_t - G(N_t) - P_{N_t}N_t - DR_t - M_t],$$

where   A   is the rate of taxation.

D   is the depletion rate.

M$_t$ is allowable depreciation. We recognize the problem of purchases of capital early or late in the period. We will assume, however, for analytical simplicity that all capital is purchased at the beginning of the period.

It should be noted that in this formulation of the present value of the firm there is no constraint imposed on the total amount of the resource available in the deposit under consideration. It follows that, in the absence of such a constraint, the marginal cost of extraction will not include the opportunity or user cost element described in Chapter 2. Given the additional effect of the implementation of the nonconventional technology on the firm's costs, the incorporation of such a constraint would greatly complicate the mathematical analysis. Furthermore, given uncertainty, the time horizon over which the innovation must prove profitable is short enough that this complicating feature may be safely ignored. For example, even though the firm may feel that (if the technology is profitable) the time horizon is 20 years, it may, in the presence of extreme uncertainty, use only a five- or ten-year horizon for evaluating the profitability of the technology.

For areas that we are considering here, only the use of nonconventional techniques can lower extraction costs sufficiently for the area to be developed, and then only under a certain set of prices of the nonconventional technology and of output. We should emphasize that the technologically optimal rate of extraction need not be the economically optimal rate. The price and the rate of discount may be such that the firm is willing to sacrifice some future production for additional current production even at a rate above that at which the total possible extraction is maximized. This may be accentuated with a very high rate of discount, such that future profits are discounted greatly.

In respect to the cost function, we will continue to make the assumption outlined in Chapter 2 concerning the effect of current and past levels of extraction on current cost. That is, we assume $\partial C_t/\partial q_{t-i} > 0$ and $\partial^2 C_t/\partial q_t \partial q_{t-i} > 0$ for $i = 0, \ldots, t$. In general, factors of production are not included in cost functions. But, because of the special problems involved here, we must include the usage of nonconventional technology in our function. We assume that in the types of areas with which we are dealing, this alternative technology will, up to a point, lower the operating cost. Clearly, there comes a point at which the diminishing returns from implementation are such that the decrease in cost from additional use does not exceed the cost of the technology. Beyond this point the firm would use no more of the alternative technology. We must also assume that implementation in one period lowers operating costs not only in that period but in some of the succeeding periods as well. Hence, we assume $\partial C_t/\partial N_{t-i} \leqq 0$ and $\partial^2 C_t/\partial N_{t-i}{}^2 \geqq 0$ for $i = 0, t$. We will make no assumptions concerning the effect of the formation characteristic, μ, on operating cost.

The firm will then select those levels of output and usage of the nonconventional technology that will maximize its net present value. Our purpose in this analysis is to examine the equilibrium and its comparative statics properties. In order to accomplish this, it will be necessary to simplify somewhat the objective function (5.1).

Let us first simplify by assuming that P_t, the price of the resource, will change at a constant rate over the period under consideration. Let this rate of change equal α. Thus, if the price at the beginning of the horizon, period zero, is P_0, in any period t, $P_t = \alpha^t P_0$. This simplifies analysis considerably, because in carrying out comparative statics we need only to examine the effect of changes in two parameters, α and P_0, rather than $(H + 1)$ parameters, P_t. We also assume that the price of the alternative technology is constant at P_N over the horizon. Since inflation is not contained in α, this term reflects only the change in relative scarcity over the period.

We will continue to assume that operating costs are a function of output and the level of nonconventional techniques employed during a period—for a given μ and a given fixed operating cost. But the cost function is assumed to change at a rate β over the horizon. If the cost function in period 0, the base year, is C_0, $C_t = \beta^t C_0$. Output and the level of nonconventional extraction in one period—say, the t-th—still affect operating cost both in that period and in succeeding periods up to some period, K, after which the effect of alternate technology in periods beyond the k-th period is zero. As in the case of α, β is assumed to be net of the effect of inflation, and therefore β reflects only relative scarcity.

We make several simplifying assumptions about the form of the operating cost function. These following assumptions are quite consistent with the empirical data that are currently available. The first assumption is that the marginal cost in any period, $\partial C_t/\partial q_t$, is a positively sloped straight line. Thus, marginal cost increases at a constant rate, which we shall denote as $W(>0)$. Second, output in any particular period raises operating cost in future periods at a constant rate, and this rate is $K(>0)$. In other words, $\partial C_i/\partial q_t = \beta^i \partial C_0/\partial q_t =$ a positive constant $= \beta^i K$ for $i > t$.

The application of nonconventional technology lowers costs in the period in which it takes place, but the decrease in costs occurs at a constant decreasing rate, designated as $F(>0)$. Thus $\partial C_t/\partial N_t < 0$ and $\partial(\partial C_t/\partial N_t)/\partial N_t =$ a positive constant $= \beta^t F$. Nonconventional technology is assumed to lower costs in succeeding periods at a constant negative rate, Z. Thus $\partial C_i/\partial N_t = \beta^i Z$ for $i > t$. But it is assumed that applications of nonconventional technology in one period will not affect the marginal productivity of an application in any other period; nor will the alternate technology affect marginal cost in any period.

Mathematically, the assumptions can be set forth as follows:

$$\frac{\partial C_t}{\partial q_t} = \beta^t \frac{\partial C_0}{\partial q_t} > 0 \text{ and } \frac{\partial^2 C_t}{\partial q_t{}^2} = \text{a positive constant} = \beta^t W. \qquad (5.2)$$

$$\frac{\partial C_i}{\partial q_t} = \beta^i \frac{\partial C_0}{\partial q_t} = \text{a positive constant} = \beta^i K \text{ for } i > t \text{ and } \frac{\partial^2 C_i}{\partial q_t{}^2} = 0. \qquad (5.3)$$

$$\frac{\partial C_t}{\partial N_t} = \beta^t \frac{\partial C_0}{\partial N_t} < 0 \text{ and } \frac{\partial^2 C_t}{\partial N_t{}^2} = \text{a positive constant} = \beta^t F. \qquad (5.4)$$

$$\frac{\partial C_i}{\partial N_t} = \beta^i \frac{\partial C_0}{\partial N_t} = \text{a negative constant} = \beta^i Z \text{ for all } i > t \text{ and } \frac{\partial^2 C_i}{\partial N_t{}^2} = 0. \qquad (5.5)$$

$$\frac{\partial^2 C_i}{\partial q_i \partial q_0} = \frac{\partial^2 C_i}{\partial N_i \partial q_j} = \frac{\partial^2 C_i}{\partial N_i \partial N_j} = 0 \text{ for all } i \neq j. \qquad (5.6)$$

$$\frac{\partial G}{\partial N_t} = \text{a positive constant, and } \frac{\partial^2 G}{\partial N_t{}^2} = 0. \qquad (5.7)$$

Incorporating these assumptions, the objective function becomes

$$PV = \sum_{t=0}^{H} \left(\frac{1}{1+r}\right)^t \{(1-L)a^t P_0 q_t - \beta^t C_0 - F_t - G(N_t) - P_N N_t -$$
$$A[(1-L)(1-D)a^t P_0 q_t - G(N_t) - \beta^t C_0 - M_t - P_N N_t]\} \qquad (5.8)$$

Equilibrium Conditions

The firm will maximize its net present value by choosing these levels of output, q, and nonconventional extraction, N, that maximize expression (5.8). Mathematically, this requires that $\partial PV / \partial q_t$ and $\partial PV / \partial N_t$ be less than or equal to zero for all values of t.* That is,

$$\left(\frac{1}{1+r}\right)^t \left\{(1 - A[1 - D])(1 - L)\alpha^t P_0 - (1 - A)\beta^t \frac{\partial C}{\partial q_t}\right\} -$$

$$(1 - A)K \sum_{i = t + 1}^{H} \beta^i \left(\frac{1}{1+r}\right)^i \leqq 0. \; t = (0, 1, 2, \ldots H) \qquad (5.9)$$

$$(1 - A)\left\{-\left(\frac{1}{1+r}\right)^t \left(P_N + \frac{\partial G}{\partial N_t} + \beta^t \frac{\partial C}{\partial N_t}\right) - Z \sum_{i = t + 1}^{H} \beta^i \left(\frac{1}{1+r}\right)^i\right\} \leqq 0.$$

$$t = (0, 1, 2, \ldots H) \qquad (5.10)$$

In equation (5.9) the term $(1 - A[1 - D])(1 - L)\alpha^t P_0$ represents net marginal revenue in the period and the term $(1 - A)\beta^t \partial C / \partial q_t$ represents marginal cost. Hence, (5.9) requires that, in any period, net marginal revenue will exceed marginal cost by the discounted marginal effect of output in the current period on costs in future periods (the last term).

In equation (5.10) the term $P_N + \partial G / \partial N_t$ represents the cost of an additional application of the nonconventional technology in that period while $\beta^t(\partial C / \partial N_t)$ represents the reduction in current cost resulting from the use of the alternative technology. Thus, equation (5.10) requires that in any period, the firm will employ the alternative technology until its discounted cost is equal to the discounted marginal reductions in cost in the current and future periods.

These results reinforce the conclusion that one would deduce intuitively. This analysis assumes, and we will continue to assume, that the equality sign holds in the first-order conditions so that the firm produces some output and uses some alternate technology in each period. Thus we assume that the present value, as set forth in (5.8), is positive.

---

*This is the first-order condition for profit-maximization. The second-order condition, necessary to guarantee a maximum, requires that the $(2H + 2)$ by $(2H + 2)$ matrix of the second derivatives of the system described by (5.9) and (5.10) (i.e., the hessian matrix) be negative definite. That is, the principal minors must alternate in sign, with the sign of the entire determinant being $(-1)^{2H + 2} > 0$. For our model, the hessian matrix is diagonal with all elements on the diagonal negative. Hence, the second-order conditions are satisfied.

Theoretical Implications and Some Comparative Statics

If one knew the marginal cost functions and the values of the parameters, one could solve for the $(2H + 2)$ decision variables, q_t and N_t—again, assuming the equality signs hold in (5.9) and (5.10). Then, if the present value is positive at these optimal rates of q_t and N_t, the firm would produce.

Theoretically we could estimate from technical data the marginal costs and marginal returns from nonconventional extraction along with the exogenously given variables r, α, P_0, etc. Next, we would estimate P_0 under the assumption of no nonconventional production. We would substitute the solutions for the N_t and q_t into the present value function, then set present value equal to zero and solve for P_N. This P_N is the *highest* price of nonconventional technology, at which present value would be barely sufficient to induce production. Next, we would artificially lower P_N in discrete steps and with the aid of the optimizing equations derive mathematically the output and usage of nonconventional techniques associated with each price of the alternative technology. We would sum the supply functions over all areas to obtain a nonconventionally produced supply.

Then, if output demand and the conventional supply were known, we would add nonconventional supply to conventional supply as we did in the graphical section to obtain an estimate of output and the usage level of nonconventional techniques in each period. Finally, we could take into consideration the effect of enforced delays of varying lengths.

Clearly, even with the simplifying assumptions, there are monumental problems in the actual solution. The primary problem of course is the absence of data for the marginal as opposed to the average cost functions. Without those data we must simplify somewhat more to carry out the empirical analysis. We can, however, carry out the comparative statics in order to determine the theoretical effect—under the assumptions set forth above—of changes in the parameters upon the choice variables. As it turns out, we can ascertain the direction of the effect and the elements that affect the magnitude of the effect, even though we cannot measure precisely the magnitude itself.

We wish specifically to determine the theoretical effect of changes in the parameters P_0, α, β, A, D, r, and P_N upon output in any period q_t, and the usage of nonconventional technology in any period N_t. To this end we assume that equations (5.9) and (5.10) hold as equalities.* We then take the total differential of the entire system to obtain

---

*Even if the inequality held in some cases, the approach described here would lead to the same results.

$$(1-A)[\phi]\begin{bmatrix} dq_0 \\ dq_1 \\ \vdots \\ dq_H \\ dN_0 \\ dN_1 \\ \vdots \\ dN_H \end{bmatrix} = (-1)\begin{bmatrix} \delta^q_{0\alpha}d\alpha + \delta^q_{0P_0}dP_0 + \delta^q_{0A}dA + \delta^q_{0D}dD + \delta^q_{0\beta}d\beta + \delta^q_{0r}dr + \delta^q_{0P_N}dP_N \\ \delta^q_{1\alpha}d\alpha + \delta^q_{1P_0}dP_0 + \delta^q_{1A}dA + \delta^q_{1D}dD + \delta^q_{1\beta}d\beta + \delta^q_{1r}dr + \delta^q_{1P_N}dP_N \\ \vdots \\ \delta^q_{H\alpha}d\alpha + \delta^q_{HP_0}dP_0 + \delta^q_{HA}dA + \delta^q_{HD}dD + \delta^q_{H\beta}d\beta + \delta^q_{Hr}dr + \delta^q_{HP_N}dP_N \\ \delta^N_{0\alpha}d\alpha + \delta^N_{0P_0}dP_0 + \delta^N_{0A}dA + \delta^N_{0D}dD + \delta^N_{0\beta}d\beta + \delta^N_{0r}dr + \delta^N_{0P_N}dP_N \\ \delta^N_{1\alpha}d\alpha + \delta^N_{1P_0}dP_0 + \delta^N_{1A}dA + \delta^N_{1D}dD + \delta^N_{1\beta}d\beta + \delta^N_{1r}dr + \delta^N_{1P_N}dP_N \\ \vdots \\ \delta^N_{H\alpha}d\alpha + \delta^N_{HP_0}dP_0 + \delta^N_{HA}dA + \delta^N_{HD}dD + \delta^N_{H\beta}d\beta + \delta^N_{Hr}dr + \delta^N_{HP_N}dP_N \end{bmatrix} \quad (5.11)$$

where ϕ is the hessian matrix of the second derivatives of the system described by (5.9) and (5.10). In this system

$$\delta^q_{t\alpha} = \frac{\partial(\partial PV/\partial q_t)}{\partial\alpha}, \; \delta^q_{t\beta} = \frac{\partial(\partial PV/\partial q_t)}{\partial\beta}, \text{ etc.}$$

and

$$\delta^N_{t\alpha} = \frac{\partial(\partial PV/\partial N_t)}{\partial\alpha}, \; \delta^N_{t\beta} = \frac{\partial(\partial PV/\partial N_t)}{\partial\beta}, \text{ etc.}$$

Employing Cramer's rule we obtain the following partial derivatives:

$$\frac{dq_t}{d\alpha} = \frac{(1-A[1-D])(1-L)t\alpha^{t-1}P_0}{\beta^t(1-A)W} > 0, \text{ for all } t > 0 \quad (5.12)$$

$$\frac{dq_t}{dP_0} = \frac{(1-A[1-D])(1-L)\alpha^t}{\beta^t(1-A)W} > 0, \quad (5.13)$$

$$\frac{dq_t}{dD} = \frac{A(1-L)\alpha^tP_0}{\beta^t(1-A)W} > 0, \quad (5.14)$$

$$\frac{dq_t}{dA} = \frac{\left\{\left(\dfrac{1}{1+r}\right)^t\left[(1-D)(1-L)\alpha^tP_0 - \beta^t\dfrac{\partial C}{\partial q_t}\right] - K\sum\limits_{i=t+1}^{H}\beta_i\left(\dfrac{1}{1+r}\right)^i\right\}}{\left(\dfrac{\beta}{1+r}\right)^tW(1-A)} < 0, \quad (5.15)$$

$$\frac{dq_t}{d\beta} = \frac{-\left[\left(\frac{1}{1+r}\right)^t (1-A)t\beta^{(t-1)}\frac{\partial C}{\partial q_t} + (1-A)K\sum_{i=t+1}^{H} i\beta^{(i-1)}\left(\frac{1}{1+r}\right)^i\right]}{\left(\frac{\beta}{1+r}\right)^t W(1-A)} < 0$$

(5.16)

$$\frac{dq_t}{dr} = \frac{t\left(\frac{1}{1+r}\right)^{(t-1)}\left(\frac{-1}{[1+r]^2}\right)\left\{(1-A[1-D])(1-L)\alpha^t P_0 - \right.}{\left(\frac{\beta}{1+r}\right)^t (1-A)W} \cdots$$

$$\frac{\left. -(1-A)\beta^t\frac{\partial C}{\partial q_t}\right\} - (1-A)K\sum_{i=t+1}^{H} \beta^i i\left(\frac{1}{1+r}\right)^{i-1}\left(\frac{-1}{[1+r]^2}\right)}{\left(\frac{\beta}{1+r}\right)^t (1-A)W} > 0 \quad (5.17)$$

$$\frac{dq_t}{dP_N} = 0 \tag{5.18}$$

$$\frac{dN_t}{d\beta} = \frac{-\left[(1-A)\left(\frac{1}{1+r}\right)^t t\beta^{(t-1)}\frac{\partial C}{\partial N_t} + (1-A)Z\sum_{i=t+1}^{H} i\beta^{(i-1)}\left(\frac{1}{1+r}\right)^i\right]}{\left(\frac{\beta}{1+r}\right)^t F(1-A)} > 0$$

(5.19)

$$\frac{dN_t}{dr} = \frac{-(1-A)\left[t\left(\frac{1}{1+r}\right)^{(t-1)}\left(\frac{-1}{[1+r]^2}\right)\left(P_N + \frac{\partial G}{\partial N_t} + \beta^t\frac{\partial C}{\partial N_t}\right) + \right.}{\left(\frac{\beta}{1+r}\right)^t F(1-A)} \cdots$$

$$\frac{\left. + \left(Z\sum_{i=t+1}^{H} \beta^i i\left(\frac{1}{1+r}\right)^{(i-1)}\left(\frac{-1}{[1+r]^2}\right)\right)\right]}{\left(\frac{\beta}{1+r}\right)^t F(1-A)} > 0 \quad (5.20)$$

$$\frac{dN_t}{dP_N} = \frac{-1}{\beta^t W} < 0 \tag{5.21}$$

$$\frac{dN_t}{dP_0} = \frac{dN_t}{dA} = \frac{dN_t}{d\alpha} = \frac{dN_t}{dD} = 0 \qquad (5.22)$$

Let us analyze equations (5.12) through (5.22) in turn. Since by assumption $0 < A < 1$, $0 < D < 1$, and $1 < L < 1$, $0 < (1 - A[1 - D])$ $(1 - L) < 1$, except for $t = 0$ the numerator of (5.12) is positive. Since W, the rate at which marginal cost changes when output changes, is by assumption positive (i.e., marginal cost increases) the denominator is also positive. Thus, $dq_t/d\alpha$ is positive. This is to be expected; if future prices are expected to increase, firms will plan to increase output.* For equation (5.13) the same reasoning applies and dq_t/dP_0 is positive. An increase in resource price increases output in each period. All terms in equation (5.14) are positive; thus, dq_t/dD is positive. An increase in the depletion allowance increases output in each period. This results because an increase in the rate of depletion results in an increased tax saving for increased output. In this way the net marginal return to production increases.

In the "typical" theory of the firm a change in the rate of taxation does not change the output decision. The firm maximizes the before-tax present value, regardless of the rate of taxation. If T is the tax rate, the firm prefers $(1 - T)$ times maximum profit to $(1 - T)$ times any other profit, no matter what the rate of taxation, as long as present value remains positive. This is not the case, however, in the situation described here. The reason for the difference is the effect of the depletion allowance on the total tax bill. If there were no depletion allowance, $D = 0$, a change in A would have no effect on output, $dq_t/dA = 0$. This can be easily seen by setting $D = 0$ in equation (5.8) then factoring out $(1 - A)$. The firm maximizes present value in the same way regardless of A. When D is positive but less than 1, equation (5.9) can be written as

$$(1 - A) \left\{ \left(\frac{1}{1+r} \right)^t \left[(1 - L)\alpha^t P_0 - \beta^t \frac{\partial C}{\partial q_t} \right] - K \sum_{i=t+1}^{H} \beta^i \left(\frac{1}{1+r} \right)^i \right\}$$
$$+ AD \left(\frac{1}{1+r} \right)^t \alpha^t P_0 = 0. \qquad (5.23)$$

Since the last term in (5.23) is clearly positive, the term inside the braces must be negative if the equality holds. Inside the braces the first term is positive,

---

*We ignore here the effect that an increase in future prices might have on H, the horizon. Recall that the horizon extends only so long as production is economical. An increase in price might extend the horizon.

while the second and third terms are negative. The only difference between the negative terms inside the braces in (5.23) and the term inside the braces in (5.15) is that $(1 - D) (1 - L)\alpha^t P_0$, while remaining positive, is smaller than the first term inside the braces in (5.23). Thus the numerator in equation (5.15) must be negative. Since the denominator is positive, $\partial q_t / \partial A$ is negative. When there is a depletion allowance, an increase (decrease) in the rate of taxation decreases (increases) output in each period. This differs from typical results. (We will return to a discussion of depletion in Chapter 11.)

Equation (5.16) is clearly negative, as would be expected. The term inside the brackets is total discounted marginal cost: the first term is present marginal cost, and the second term is future marginal cost. Since the first term must be positive (nonnegative if $t = 0$) and the second term nonnegative, the numerator of (5.16) is positive. With the positive denominator, $dq_t / d\beta$ must be negative. As would be expected, an increase in the expected rate of change of operating cost decreases output in each period.

In the "typical" model a change in the discount rate has no effect upon the decision variables. Alternatively, in the model used here present production effects are discounted by a different amount. The mathematical logic that enables us to sign $\partial q_t / \partial r < 0$ is quite complex, even though the economics is not.

First, note that we can write the numerator of (5.17) as

$$\left(\frac{1}{1+r}\right)\left\{-t\left[\left(\frac{1}{1+r}\right)^t(1-A[1-D])(1-L)\alpha^t P_0 - (1-A)\beta^t\frac{\partial C}{\partial q_t}\right]\right.$$

$$\left. -\left[(1-A)\sum_{i=t+1}^{H}(-i)\left(\frac{1}{1+r}\right)^i\beta^i\right]\right\}. \tag{5.24}$$

Except for the $(-t)$ times the first term in brackets and the $(-i)$ times each term in the summation making up the last term, equation (5.24) would equal equation (5.9) and would therefore be equal to zero. Since the last term in equation (5.9) is negative, the first term in braces must be positive. Therefore, $(-t)$ times the first term is negative, and $(-i)$ times every element in the last term makes the last term positive. Since $t < i$ $(i = t + 1, \ldots, H)$ the negative portion of equation (5.24)—$(-t)$ times the term in brackets—is algebraically less than (absolutely greater than) $(-i)$ times each element in the last term. Thus, equation (5.24) is positive. In other words, let $[\phi]$ stand for the first term in brackets in equation (5.24) and $[\psi]$ stand for the second term in brackets, and let there be only one element in the summation of the second term. From equation (5.9)

$$[\phi] = [\psi]$$

Since $t < i$,

$$-t[\phi] > -i[\psi].$$

Thus, $\left(\dfrac{1}{1+r}\right)\left\{-t[\phi]+i[\psi]\right\} > 0,$

which means from (5.17) that $dq_t/\partial r > 0$. An increase in the rate of discount increases output in each period except period H, in which there is zero effect.

Equation (5.18), in which $dq_t/dP_N = 0$, presents a rather strange phenomenon. In most models of the firm, a change in the price of one of the inputs—in this case nonconventional technology—changes output. P_N, however, has no effect upon output, because the first $(H+1)$ equations determine output in all periods. Therefore, the price of the nonconventional technology has no effect upon output. Only under the assumption that the alternative technology has no effect on *marginal* (as opposed to average) costs will this result hold. After all, it is marginal conditions that determine output. Obviously, if P_N increases so much that present value becomes negative, output falls to zero.

Let us turn now to the effect of parametric changes upon the use of the nonconventional technology. Skipping ahead, let us first dispose of equation (5.22). Changes in α, P_0, and D have no effect upon the usage of nonconventional technology in any period for the same reason given above that $dq_t/dP_N = 0$. These variables do not enter the last $(H+1)$ equations in (5.10), and therefore have no effect upon N_t. There is no interdependence between the equations that determine q_t and those that determine N_t. Firms use the alternative technology until the reduction in all operating costs equals the total cost of an additional application. Those variables that exert their effect only upon revenue—such as α, P_0, and D—have no effect upon these marginal conditions. A, the rate of taxation, does enter into the equation that determines N_t. But A has no effect upon the marginal conditions; that is, the last $(H+1)$ equations show that $(1-A)$, a positive number, times the marginal conditions equals zero. Thus, no matter what the value of A, the same values lead to the final equilibrium conditions in the case of N_t. Thus $dN_t/dA = 0$.

The sign of equation (5.19), $dN_t/d\beta > 0$, follows directly from the assumptions. Since $\partial C_i/\partial N_t < 0$, the first term within the brackets is negative for $t > 0$ and zero for $t = 0$. Since $Z = \partial C_i/\partial N_t < 0$ for $t < i \leq H$, the second term, of numerator is negative. Given that the denominator is positive, $dN_t/d\beta < 0$. This would be expected. Since $\beta' \partial C/\partial N_t$ is the marginal reduction in cost from nonconventional technology, a decrease in β reduces the marginal return to the alternative technology and, therefore, the usage of nonconventional extraction techniques is reduced.

Following the same line of reasoning used in the case of dq_t/dr, we can write the numerator of (5.20) as

$$- (1 - A)\left\{ t \left(\frac{1}{1+r}\right)^t \left(P_N + \frac{\partial G}{\partial N_t} + \beta^t \frac{\partial C}{\partial N_t} \right) + Z \sum_{i=t+1}^{H} \beta^i \left(\frac{1}{1+r}\right)^i \right\}.$$

$$(5.25)$$

From (5.10), except for t times the first term and i times each element in the second term, equation (5.25) would equal zero. Since $Z \sum_{i=t+1}^{H} i\beta^i (1/[1+r])^i$ is negative, $(1/[1+r])^t (P_N + \partial G/\partial N_t + \beta^t \partial C/\partial N)$ is positive. Therefore, the term inside the braces is negative because $t < i$ and the numerator is positive. Thus an increase in r increases use of nonconventional extraction in each period. Finally, it is obvious both from equation (5.21) and intuitively that the use of the alternative technology varies inversely with its price. This is similar to the results derived for all types of inputs in production theory.

A Brief Overview of the Mathematical Analysis

As in Chapter 2, we specified the objective of the firm to be the maximization of its present value (i.e., net worth). We again incorporate the dynamic feature that current output affects not only current operating costs but also future costs; thus, the marginal cost of current extraction includes the effect not only on current costs but also on future costs. We did not, however, include any constraint on the amount of the resource available in the deposit. Therefore, the opportunity or user cost element of marginal cost was not considered.

After employing some simplifying assumptions to make the mathematics tractable, the present value of the deposit was specified to be a function of the firm's decision variables, output and the level of use of the nonconventional technology, and some exogenous parameters: the rate of change in output price over time, the rate of change in cost over time, the rate of income taxation, the allowed rate of percentage depletion, the rate of discount (i.e., the rate of interest), and the price of the nonconventional technology. From this model, it could be seen that the profit-maximizing firm would employ the nonconventional extraction technique until the discounted marginal reductions in current and future operating costs are equal to its discounted cost.

We then considered the effects of changes in the exogenous parameters on the profit-maximizing values of output and the level of wage of the nonconventional technology. With respect to output, an increase in the rate of increase of output prices, the rate of percentage depletion, or the discount rate would increase the rate of current extraction, while increases in the rate of income taxation or the rate of increase in operating costs would reduce it. It was found that an increase in the rate of increase in operating costs or the discount rate would increase the level of use of the nonconventional technology, while an increase in the price of the nonconventional technology would reduce its use.

POTENTIAL APPLICATIONS

We have thus far concentrated solely upon the theoretical analysis of the economic feasibility of new nonconventional technologies in mineral extraction. Now that we have set forth the theory, we are in a position to examine how one can actually use the structure to answer certain interesting empirical questions—for example, what would the price of the new technology or the resource have to be to make that new technology economically feasible? Or what would the discount rate have to be with given resource and technology prices?

If we had marginal cost data for a specific nonconventional extraction technique, such as deep-seabed mining or the use of nuclear devices to stimulate gas production or in the leaching of copper, it would be a straightforward, albeit mathematically complex, matter to estimate the variables discussed in the previous section. The problem is that the usage of most nonconventional techniques is so new and the technology so uncertain that marginal functions are simply not available. Generally, only a few preliminary experiments will have been conducted, and these few experiments would provide the only source of data.

Nonetheless, the techniques described in the previous section do permit feasibility analysis given the availability of ranges of estimates of some of the necessary data. In this section we first describe the basic methodology and needed data and then outline the methodology for several specific experiments to determine the economic feasibility of nonconventional technologies. We will begin with a brief outline of the basic methodology used.

Basic Methodology

In Chapter 2 and in the previous section we showed that the firm would attempt to maximize the present value of a given field or deposit over its time horizon. As noted, data unavailability prevents our using the maximizing conditions to predict conditions under which a nonconventional technology would be economically feasible. There is, however, an alternative method that can frequently use data from preliminary experiments in order to make preliminary feasibility estimates.

To begin the analysis, consider the following conceptual experiment. Suppose it would not be profitable to exploit a certain field or deposit with any conventional technology. But assume that at some price of a specific nonconventional technology the field or deposit would have a positive present value. Let the present value function be that set forth in equation (5.8) above, and let all values of data—such as prices, costs, discount rates, taxes, etc.—be known. In fact, let everything except the price of the nonconventional technology be known.

It is probable that the experimental data are in the form of fixed-proportions production functions. If nonconventional techniques are used, resources can be extracted at some given rate using a given level of the alternative technology. Under this method, operating cost is a given amount, as is capital cost. We must assume, of course, that these figures are derived using some form of maximizing techniques like those set forth above.

We can assume one thing about the firm: It would not exploit the field or deposit if the price of the nonconventional technology were so high that the present value—including the return to capital—would be negative. That is, the price of the nonconventional technology must be low enough to make present value zero or above. Thus, the *maximum* price at which the nonconventional technology would be used would be that price at which the present value is zero. Thus, if we had estimates of all other variables, we could use these estimates in our present value function (in this case equation [5.8]), set the function equal to zero, and solve for the price of the nonconventional technology. This would be the maximum price at which that technology would be feasible. Any lower price of the technology would, of course, give a positive present value.

Alternatively, if we had estimates of all data but the price of the resource, including the price of the new technology, we could set present value equal to zero and solve for the *minimum* resource price that would make the technology economically feasible. Any higher price would naturally give a positive present value. Or we could use the same techniques to find the maximum discount rate that would make the new technology feasible. As we shall see, however, we will probably never have specific estimates of the variables, including costs, but only ranges of estimates. Thus, using ranges of estimates of the variables, we will be able to obtain only ranges of estimates of the prices, costs, discount rates, etc. The next subsection will show how such estimates are carried out. First let us show what data are needed. These are the variables used in the present value function, equation (5.8).

For each type of application under consideration here we would require certain technical and certain economic data. The technical data are as follows:

1. Capital costs (F_t). As noted above these are capital additions less depletion and depreciation.

2. Operating costs (C_t), net of the cost of the nonconventional technology $(P_N N_t)$ and any hazard costs. These are the total yearly operating costs at the optimal rate of extraction.
3. Yearly output per application of the new technology (q_t). This is assumed to be derived using maximizing techniques.
4. The yearly hazard cost $G(N)$, if any.
5. Land use rate (L).
6. Tax rate (A).
7. Rate of depletion allowance (D).
8. Depreciation in the t-th period, M_t.

We could obtain *ranges of estimation* for the following economic data:

1. The expected price of the resource in each year for the period under consideration. We use only the estimate of P_0, then assume that in any year the price is $\alpha^t P_0$, where t is the year and α is the estimate of the rate of change in resource price net of inflation. We obtain the value of α from the price projections for the relevant resource.
2. The rate at which operating costs are expected to change over the relevant period, β. We use the operating cost estimate for the first period, then assume that cost in period t is $\beta^t C_0$, where C_0 is the estimate for the first period.
3. The price of nonconventional technology, P_N, in each period.
4. The going rate of return on capital, r, net of taxation.
5. Regulatory delay times expected.
6. The time horizon for the project, H.

Estimations of Prices and Rates of Return

Our first task is to find the *highest* price of the nonconventional technology that would make it economically feasible under several sets of circumstances, where economically feasible means that under the optimizing conditions used to determine the rates of output, the present value is nonnegative.

Let us assume first that there are no legally imposed delays. We proceed in the following way. Use the best available estimates for r (the rate of discount), α (the rate of change in resource price), and β (the rate of change in cost). Assume that the cost variables are known. Take the highest estimate for the resource price, P_0, assuming that this price is freely determined in the market. Set present value in equation (5.8) equal to zero, and solve for the price of the nonconventional technology, P_N. This is the highest price that would occasion any usage under these assumptions. Obviously, any lower price would increase present value to a positive amount under these conditions. Repeat the experiment using successively the best available and lowest estimates of P_0. Repeat the entire experiment for the lowest and the highest estimates of r, the discount rate.

Thus we have the following estimation matrix for α_b and β_b (best estimates) when there are three estimates of P_0 and r.

	P_l	P_b	P_h
r_H	P_N^{lh}	P_N^{hb}	P_N^{hh}
r_b	r_N^{bl}	P_N^{bb}	P_N^{bh}
r_l	P_N^{ll}	P_N^{lb}	P_N^{lh}

Next, one could, if one so desired, carry out the same experiments under the high estimates and low estimates for α and β. We would then have three matrices, each of which has nine estimates P_N. Assuming equal probability for each of the 27 estimates, we then obtain a frequency histogram for the *highest* price that makes an application *economically*, as opposed to technically, feasible.

Thus far we have assumed no legally imposed delay. But there is certainly some probability of legal delays in production. That is, legal action may cause production to be ceased for the first period, the first two periods, and so on. Therefore, in these production periods q, N, C, and G(N) will be zero. Thus we can estimate maximum feasible prices under the assumption of not only zero delay but also positive delay periods. We could first assume a one-period delay time. Therefore, we will set q_0, and N_0, C_0, and $G(N_0)$, equal to zero. Keep F_0 and all other variables the same as above and repeat the experiment already described. That is, we solve for the highest price of the nonconventional technology that would make present value nonnegative under all the different conditions described for the above experiment, but we set production and all related variables in period zero equal to zero. We could repeat the experiment with a two-year delay, a three-year delay, and so on, up to the maximum shown by delay studies. This series of experiments would show the *highest* price of the technology that would make nonconventional extraction economically feasible under all circumstances.

Next we might wish to estimate the minimum resource price that would make nonconventional extraction economical under several sets of assumptions. To this end, use the best estimates of α, β, and r. Using the high, the best, and the low estimates of P_N, the price of the nonconventional techniques, set present value equal to zero and solve for P_0. This result indicates the lowest resource price that would make production feasible under the given circumstances. Performing the same experiment under the assumption of high and low estimates of r, we would obtain the estimation matrix:

	P_N^l	P_N^b	P_N^h
r_h	P_0^{hh}	P_0^{bb}	P_0^{hh}
r_b	P_0^{bl}	P_0^{bb}	P_0^{bh}
r_l	P_0^{ll}	P_0^{lb}	P_0^{lh}

Repeat the experiment under the assumption of all conceivable numbers of delay times. Finally, do exactly the same thing for high and low estimates of α and β. Thus, we obtain a range of estimates for resource prices and the conditions under which this range is attained.

We could then estimate the range of rates of return—net of taxation—that would make nonconventional extraction economical under different sets of conditions. First set α and β equal to the best estimates. Then combine the high, low, and best estimate of the price of the technology with the high, low, and best estimates of the price of the resource in question. Use all nine combinations of P_0 and P_N in the present value function, which is set equal to zero. Solve for the net rates of return that yield zero present value. Choose the highest; this rate is the highest rate of return that yields a nonnegative present value under each set of circumstances. We obtain the matrix:

	P_l	P_b	P_h
P_N^h	r_{hb}	r_{hb}	r_{hh}
P_N^b	r_{bl}	r_{bb}	r_{bh}
P_N^l	r_{ll}	r_{lb}	r_{ll}

We repeat the experiment for delays of one through five years and for the low and best estimates of α and β under all conditions. In this way we obtain a range of estimates for an economical rate of return and the conditions under which this range is attained.

Estimation of the Effect of Nonconventional Extraction upon Total Consumption and the Demand for Nonconventional Techniques

Thus far we have analyzed only the economic feasibility of nonconventional extraction at the projected market price of the resource. Now let us analyze the economics under the possibly more realistic assumption that the additional supply will lower the market price of the resource.

A complete graphical analysis should aid significantly in understanding the technique of estimation to be used. Assume that we wish to derive the demand for a specific nonconventional technique in the zero period. In Figure 5.6, panel A, let DD be the demand for the resource in question and let SS be the supply function under the assumption of only conventionally produced output. P_0 is the expected price under the assumption of no augmented production.

Under ideal circumstances we would derive the demand for nonconventional techniques as follows. With several prices of the technology we would determine the nonconventional supply associated with each price and add each new supply curve to the conventional supply. Under the assumption of decreasing marginal returns, the nonconventional supplies would be upward sloping, reflecting increasing costs as nonconventional production increases. Three of these augmented supplies are GH, CF, and AB, which are associated with three distinct prices of the nonconventional technology. The total supply would be SGH, SCF, or SAB. If the price of the technology is so high that nonconventional supply is everywhere above P_0, there would be no augmented production, and no nonconventional technology would be employed.

To estimate the demand for the nonconventional technology we would determine *total* output at each new equilibrium price. The difference between conventional supply and demand at the new equilibrium price is the output that is economically possible using nonconventional extraction. The production function gives the level of use of the nonconventional technology that would be associated with a specific quantity of production demanded in this period. Summing over all types of applications, one would obtain the total demand during the entire period. This procedure is not shown in Figure 5.6.

The data, however, are not such that we can use this technique exactly as described. As noted above, experimental data generally indicate a fixed-proportions production function. Thus we must improvise slightly to get an acceptable estimate of the demand for a nonconventional technology.

Assume once again in Figure 5.6 that SS is conventional supply; P_0 and Q_0 would be price and output during period zero in the absence of any augmented production. Next we simply assume that the price of the resource is arbitrarily set at P_0^1. At P_0^1 quantity demanded would be Q_D^1, and quantity conventionally supplied, Q_S^1. Thus, estimated excess demand at P_0^1 is $Q_S^1 Q_D^1$.

Now we wish to compute the price of the nonconventional technology that would lead to a horizontal supply at P_0^1. Recall that our assumption of fixed-proportions production leads to a horizontal supply. We then substitute P_0^1 into the present value function, using the best estimate of α, β, and r. Setting present value equal to zero, we solve for P_N in order to obtain the

FIGURE 5.6

The Demand for Nonconventional Technology

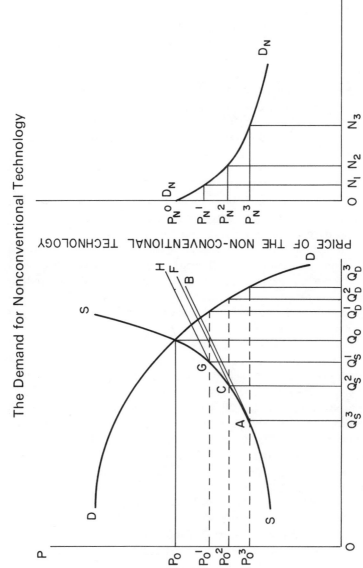

(A)

(B)

highest price of the technology that would make production economical under the best estimates of α, β, and r when the resource price is P_0^1. Assume that this price is P_N^1 in panel B of Figure 5.6. Finally, from the production figures we obtain the level of use of the nonconventional technology associated with output $Q_S^1 Q_D^1$, the former excess demand at P_0^1. That is, with the augmented production, when the price of the technology is P_N^1, equilibrium occurs at P_0^1, Q_D^1. If the level N_1 is associated with output $Q_S^1 Q_D^1$, we obtain a point on the demand curve in the zero period under the given set of assumptions. Next, let the resource price be P_0^2. Substitute P_0^2 and the best estimates of α, β, and r into present value, then set present value equal to zero. We obtain a price of, say, P_N^2. This is the highest price of the nonconventional technology that would make this type of production economical at a resource price of P_0^2. Thus, in Figure 5.6B P_N^2, N_2 is another point on the demand for nonconventional technology. Setting resource price at P_0^3 we obtain the point P_N^3, N_3 in the same way. We choose other arbitrary prices and obtain the demand $D_N D_N$ in panel B. We do the same thing for each period to obtain the total demand.

It would be well to emphasize here, before going on, three important general points made only implicitly in the above graphical analysis. First, each P_N—the price of the nonconventional technology—for the demand function shown in Figure 5.6B is the *highest* price that would be associated with each horizontal, nonconventionally produced supply of the resource shown in Figure 5.6A. That is, we set the price of the resource in period zero equal to some arbitrary value, say, P_0^1. Then we set present value equal to zero (although r is positive) when solving for P_N. Obviously, at some lower P_N, the present value could be positive at P_0^1. In any case, this value would probably be a rather good approximation, if r is a good measure of the going rate of return. Furthermore, the estimates depend upon the assumptions made about the values of α and β.

Second, the level of the nonconventional technology associated with each price of the technology is possibly biased upward if increasing costs actually prevail, because in our estimations we use the assumptions of constant costs of production in each period. Therefore, the estimated level of use demanded at each price is the *largest* that would be demanded under the given set of circumstances. Suppose price is set at P_0^1 and we find the associated P_N using the methods set forth above. Using constant costs we assume that output can expand without increasing average costs at all—that is, we do not assume decreasing returns. In this situation, with constant costs, augmented production can and would make up the entire excess demand of $Q_S^1 Q_D^1$.

But, if decreasing returns are encountered, costs may rise as augmented production increases. Perhaps in Figure 5.6A GH is the true supply. Thus at the given P_N—deduced here as P_N^1 in Figure 5.6B—less than $Q_S^1 Q_D^1$ would be added to conventional supply. In addition, the resource price would have to rise. The level of use associated with this decreased augmented supply would be less than the level associated with $Q_S^1 Q_D^1$. Thus, our estimate that N_1 is associated with P_N^1 is, in this case, a slight overestimate. If, as in the example GH, costs rise only a little, the level of use estimated is a good estimate of the level that would actually be demanded at P_N^1. Alternatively, production costs may increase rather rapidly after a point. In fact, the augmented supply can become almost vertical, indicating a possible limit to augmented production. The level of usage associated with the reduced output may be a little lower than the level associated with $Q_S^1 Q_D^1$, depending on the slope of supply. To summarize, the demand for nonconventional technology shown in Figure 5.6B indicates the *largest* level that would be demanded at a price no higher than that given on the vertical axis. It is a good approximation if costs do not increase much as nonconventionally produced output is expanded. Note also that demand is derived using best estimates of α, β, and r. Different estimates would occasion different demands.

Third, let us emphasize that the method of estimation described here is not simply an extension of the "gapsmanship" approach criticized above. Here economic factors—not some externally imposed conditions, such as how much one thinks society will need—determine the amount of nonconventional production. The estimate describes the solution the market would reach at each given price of the nonconventional technology, assuming from the data that it is used in fixed proportions to output. There is no "fill the gap" or "need" approach. We deduce total supply under the given set of economic conditions. The market determines a new equilibrium solution, and the production function determines the level of use demanded under the circumstances. The closer the assumptions are to actual conditions, costs, rates of increase, etc., the more accurate the estimate.

In actually carrying out the estimations we would begin with the best estimates of P_0, price in period zero, α, β, and r, and then let the resource price, P_0, fall in discrete steps. In reality we would not know the exact forms of supply and demand. Thus, we must estimate excess demand. Using the best estimates of supply and demand elasticities (the responsiveness of quantity supplied with conventional techniques and the responsiveness of quantity demanded to changes in the resource price), we would compute the extent to which quantity demanded would exceed quantity supplied at each arbitrary price in each period.

Using the technological data, we estimate the level of nonconventional technology associated through the fixed proportions production function with each level of excess demand. This is the level necessary to augment conventional supply sufficiently to satisfy demand in each period. We substitute each estimate of resource price ($\alpha^t P_0$) with the best estimates of β and r in the present value function and solve for the price of the technology that leads to zero present value under the above conditions. As noted above, this result is the *highest* price associated with each resource price below equilibrium in each period.

Under the specified set of assumptions the augmented (non-conventional) supply at each of the prices of the nonconventional technology derived when present value is set equal to zero is the supply that would be forthcoming when the price of the technology is no higher than that set forth in the solution. This is a market-determined supply, not a supply to "fill the gap." According to the best estimation there is a given level of use associated with each augmented level of output. Assuming that nonconventional production fills the entire excess demand, the amount of excess demand determines the level of usage associated with the given price.

Thus far, we have only talked about estimating the demand in period zero. We must carry out the same type of approach, therefore—after setting present value equal to zero and solving for P_N—for all periods beyond the first. We use estimated conventional supply-and-demand elasticities to obtain an estimate of excess demand at each arbitrarily determined price. We repeat the process for all periods and sum in order to find the total demand over the entire period.

Next, we would carry out precisely the same calculations in order to obtain similar results for the highest estimate of P_0 and with the higher estimates of demand-and-supply elasticity, holding α, β, and r at the best estimates. This allows us to obtain a different demand under alternative circumstances. We do the experiment with the lowest estimate of P_0 and both elasticities. Since calculations with other estimates of α and β would be a little redundant, we would not vary these terms. Finally, we would do the experiment under the assumption of delays.

SUMMARY

In a sense this chapter has been a bit of a digression from our main line of analysis. Until Chapter 5, with a few exceptions, we have analyzed the firm under the assumption of a given technology. And we shall generally continue to use this approach. But because of the increasingly persistent talk and writing about new technologies replacing old in mineral extraction, about nonconventional extraction procedures augmenting conventionally extracted supply, and about new sources of energy replacing old, we thought a brief digression into the economics of such changes would be important, particularly in view of the poor economics thus far used to analyze the feasibility of new technology. We did not and will not attempt to examine empirically the feasibility of any specific nonconventional technology for any specific mineral. Our major objective was simply to provide a basic framework of analysis for future feasibility studies. These studies could be simpler or more complex than those described here. That is why we attempted throughout the chapter to carry out the analysis as generally as possible. Thus, the framework set forth here can easily be adapted for specific cases.

To summarize, in this chapter we have examined technological innovation in the extractive firm to consider the conditions under which nonconventional extraction techniques will be employed, the effect of these alternative techniques on the market price of the resource, the equilibrium-use levels of these nonconventional techniques in a profit-maximizing firm, and the effects of changes in variables (e.g., the interest rate) on the use of nonconventional techniques. As we pointed out, much of the preceding economic analysis was at best misleading. This is due to the fact that most of those who did the analysis attempted to determine whether nonconventional technology could fill some production "gap," and they failed to consider the effect of changing prices on both the supply of and demand for the resource.

We first employed a simple graphical approach to examine the issue. We explicitly considered the effect of changes in the market price of the resource on both supply and demand and the effect of higher prices for the nonconventional extraction techniques on the supply function. From this analysis, our most significant conclusion was that in a freely functioning market nonconventional extraction techniques will lower resource prices rather than raise them—contrary to what was presumed by most previous economic analyses.

We then employed a more rigorous mathematical approach to examine the reaction of a profit-maximizing firm. In this analysis we extended the model developed in Chapter 2 to incorporate the effects of nonconventional technology. Nonetheless, we continued to consider the objective of the firm to be the maximization of its net present value. We did not, however, impose any constraint on the total amount of the resource available in the deposit; so marginal cost will not include the opportunity or user cost described in Chapter 2. From this analysis it was found that in any period the firm will employ nonconventional extraction until its discounted cost is equal to the discounted marginal reductions in operating cost in the current and future periods. Finally, we considered the effects of parametric changes on the profit-maximizing level of usage of nonconventional extraction techniques. It was found that an increase in the expected rate of increase in operating expenses or the interest rate (the rate of discount) would increase the level of use of the nonconventional techniques, while an increase in the price of the nonconventional technology would reduce its use. The level of use of the nonconventional technology was found to be independent of the rate of increase of future prices of the resource, the depletion allowance, and the rate of income taxation.

As noted above, within the limits data availability, our methodology could be employed to examine a specific nonconventional extraction technique, such as deep-seabed mining or the use of nuclear devices to stimulate gas production or in the leaching of copper. Using available data and a range of estimates about future prices and costs, one could use our model to obtain estimates of the economic feasibility of the technique. In particular, one could estimate the highest price of the technology for which it would be economically feasible, the minimum price of the resource that would make the nonconventional economically feasible, the rates of return under which the technology is feasible, and the effect of nonconventional extraction upon total consumption and the demand for nonconventional techniques.

PART II

THE IMPACT OF GOVERNMENTAL POLICIES ON THE MINERAL-EXTRACTING SECTOR

6

An Introduction to
Environmental Controls

In Chapters 7 through 9 we set forth the theory of environmental controls and taxation in mineral extraction, present an empirical analysis of the economic impact of environmental controls and taxation, and analyse the calculation of cost of and benefits from the imposition of environmental controls in mineral industries. While taxation of mineral-extracting industries is a rather straightforward concept, the fundamentals of environmental controls are not so straightforward because of several complicating issues. Therefore, in this chapter we shall introduce the reader to some of these complicating issues and to the basic concepts behind environmental controls. This chapter serves as an introduction to much of the material in the next three chapters. Since it is primarily an introduction, the material is designed for the readers who are not economists. Therefore, this chapter is not designed for and should be omitted by professional economists, because most of it is basic economic theory.

Environmental controls in general take the form of externally imposed constraints upon firms and industries. The purpose of the constraint is to reduce or at times even to eliminate completely the amount of pollution or by-product that the firm or the industry imposes upon the environment. The pollution or the by-product is a cost to either society as a whole or to some segment of society. The controls or restrictions are designed to reduce or eliminate these "social costs" (called "externalities") that are not borne by the firm.

Prior to the imposition of controls, firms are generally assumed to act as though the pollution is costless to the firm. The constraints, in effect, are an attempt to internalize to some extent these external costs. Thus, controls have an important impact, because they increase the costs of the individual firms that are being constrained. Either firms reduce their production—and society is deprived of the lost output—or firms respond to the controls by increasing their resource use to reduce pollution, or perhaps they do both. Therefore, the constraints may cause a reduction in output and profit, an increase in costs, and either an increase or a decrease in employment and investment. In any case the costs are, in the long run, imposed upon society in the form of decreased output.

The impression sometimes given in popular literature dealing with pollution and environmental controls is that societies should opt for zero pollution, and that if a society does not desire zero pollution (as has been the case in every society to the present), then zero pollution should be externally imposed by those who are wiser than the other members of society. This approach is obviously nonsense. Pollution reduction has a cost; the cost is the goods and services foregone by society. Elimination of *all* pollution would have almost an infinite cost to society.

But in the absence of enforceable property rights in air, water, etc., society may wish to reduce pollution below the level that would obtain in the absence of all external restrictions. The question is how much reduction is optimal from a social point of view, remembering not only that pollution reduction is desired by society but also that society must give up goods and services in order to reduce pollution. Thus, society must balance the benefits from the reduction of pollution with the cost of the reduction in goods and services.

In order to introduce environmental controls, we begin with a very brief discussion of the topic of externalities in general. In this section we show the major issues involved and point out that environmental controls are only one method by which externalities may be controlled. We then provide a simplified analysis of environmental controls, which forms the basis for an examination of the effects of the controls in the mineral-extracting industry to be set forth in Chapters 7 and 8. We conclude the chapter with a brief introduction to the analysis of the benefits of such controls, a topic to be addressed in more detail in Chapter 9.

EXTERNALITIES: AN OVERVIEW[1]

An externality is said to exist when the activity of one economic unit either damages or benefits another. More specifically, an externality means that the activity of one unit alters the cost conditions facing another; thus, for an activity imposing externalities, the marginal *private* cost or benefit does not equal the marginal *social* cost or benefit. While the issue of externalities and their control has received a great deal of attention recently, the topic has a long historical tradition beginning with Alfred Marshall and A.C. Pigou.[2] In this section we will not try to provide a survey of the literature; rather, we wish to provide an overview of the discussion.[3]

An external economy is said to exist when marginal social cost is less than marginal private cost. Thus, when marginal private cost equals marginal social benefit, marginal social cost is less than marginal social benefit. More resources should be allocated to producing the commodity in question, but they are not. On the other hand, an external diseconomy exists when marginal social cost exceeds marginal private cost. At such a point, marginal social benefit is less than marginal social cost. An undesirably large amount of resources is allocated to producing the commodity in question.

At this stage it is quite reasonable to ask how marginal private cost and marginal social cost can diverge. One of the chief answers is "by the existence of ownership externalities." Briefly, this means that there is some scarce resource owned by a person, but for some reason the owner cannot charge a price for the use of this resource. And when prices cannot be charged, misallocation of resources results.

Our discussion up to this point may seem to be a bit murky. Perhaps we can simplify somewhat with a few analytical examples. We turn first to a problem concerning urban renewal.[4]

Based on economic theory we would expect that an individual who owns a piece of property would keep the property developed and repaired so long as the marginal benefits exceed the individual marginal costs of such repairs. There should be no reason for "urban blight" under these circumstances. But with certain ownership externalities, completely rational people may well allow their property to deteriorate.

Note first that the value of a piece of urban property, houses, apartments, and so on depends to some extent on the condition of other property in the neighborhood. Suppose there are only two properties, one owned by owner A, the other by owner B. Each is attempting to decide whether to make an added investment in repair. Both are reaping some return from the property. Each owner has made an initial investment and has an additional sum invested in bonds. Each is making an average return of 4 percent from the property and the bonds, and this return is expected to continue even if no money is taken out of bonds and put into property repair.

If both owners take their money out of bonds and invest in property repair, each will make a return of 7 percent. Clearly each will be better off. There is a problem, however. Suppose one owner invests the bond money in repairs while the other does not. One property is improved and the other remains run down. The owner who improves the property gives up the return from the bonds, but the property return does not increase much because it remains next to a deteriorated property. In this case the improving owner's total return falls to 3 percent. The loss of bond income more than offsets the increase in income from improving the property.

On the other hand, the owner who did not improve the property retains the bond income and in addition finds that the return on the unimproved property increases because of being next to an improved piece of property. Perhaps the total return from property and bonds increases to 10 percent. These changes in return result no matter who does the repair and who does not. The owner who does not repair the property keeps the bond return and benefits from the better neighborhood.

Suppose both owners know the expected return but do not know what the other will do. If owner A decides to invest it will be to owner B's advantage not to invest: a 10-percent compared to a 7-percent return. Owner A knows this and therefore knows that the investment will cause a decrease in total return from 4 to 3 percent. Owner B knows the same conditions apply to his own investment decision also. Therefore, neither invests. However, if both would invest, each would be better off.

The above problem involved ownership externalities. Because each owner imposes external economies on the other, fewer resources are allocated to improvements than would be in the absence of the externality. In that case, if one person could acquire the property rights to the entire area, all could be made better off. But the more owners that are involved, the more difficult it becomes to internalize the externalities. Other similar cases involve ownership externalities in production and property rights. The following is an example of a common-property problem.[5]

Petroleum technology is such that for a given pool of oil there is a particular rate of extraction that maximizes the total amount of oil that can be extracted. This rate, to be sure, is the technically optimal rate and not necessarily the economically optimal rate. But suppose that for a given pool the technically optimal rate is the same as the economically optimal rate. Suppose that the pool of oil is so large that many people own pieces of land over the pool, and therefore, many people can pump oil from the pool. Suppose also that the amount pumped from this particular pool does not affect the world price of crude oil.

If one person owned all of the land over the pool, oil would be pumped at the profit-maximizing optimal rate. But if many people can pump from the pool, there is no incentive to pump at the optimal rate. Any single extractor would have no incentive to cut back. In fact, each landowner would have the incentive to pump as rapidly as possible in order to get as much as possible before the other landowners get it all. If one person cuts back on extraction, it simply means others get more. Thus, oil is pumped at a rate that is greater than is economically optimal. In this case, because the owners are imposing external diseconomies on one another, more resources are allocated to current production than would be optimal, and the pool is depleted too rapidly.

Up to now in the discussion of externalities the property rights of resource owners were fairly well defined. Problems would arise only because the production or investment or even consumption decisions of some affected the incomes or utilities of others. The problem of externalities becomes more serious when property rights are not well defined or when no one has property rights, as in the case of some scarce resources.[6]

The most complete concept of property rights is that of ownership of one's property. One can use the property in any way, subject to laws concerning injury to other parties. A less complete right is the right to use the property of someone else and to gain benefits from its use, but not to sell it or alter its form. Most rental properties and community-owned properties fall into this category. Finally there is the right to hold a good but not to use it, change it, or sell it.

As it turns out, most of the problems of externalities result from two situations. The first is incomplete or communal assignment of property rights or even no assignment of property rights. The second problem results when certain uses of one's fully owned property have harmful effects on someone else's property.

A scarce resource that is not owned is both overused and underproduced. No early settlers in the United States had the incentive to postpone chopping down trees or to plant forests. If one group did not do it, others would. Forests were destroyed. Buffalo, which were not owned, were practically wiped out, while cattle, which were owned, were not. Rivers, which were not owned were polluted. No one owned the valuable whales, and they practically disappeared when whale oil was the major source of light and lubrication. Neither competition nor human greed was the problem. The problem was that no one had the incentive to kill fewer whales so that the whales could reproduce at a rate sufficient to maintain the population. Similarly, no one has the incentive to produce goods that cannot be owned and hence from which returns cannot be realized. There is no point in individuals planting trees on publicly owned land. No one can reap the economic benefits from keeping publicly owned beaches clean. This is not to say that some people will not refrain from littering out of civic-mindedness; there is no economic incentive to do so, however, if there is no cost. People do not normally throw beer cans in their own yards, but many do so along the roadside.

As we have implied, publicly owned property gives rise to similar problems as nonowned property. A government- or community-owned property may not be used efficiently. Suppose a community owns a large piece of property that is better suited for growing vegetables than for cattle grazing. If anyone can use the property, it will probably be used for cattle grazing, because if anyone can harvest the vegetables, growers will have to expend added resources to protect their crops. Cattle owners can drive their cows home at night from the community property.

One of the fundamental problems of economics is the legal assignment of property rights when the marginal social cost of some activity exceeds the private social cost. A factory or group of factories pollutes a publicly owned river; the owners of property along the river downstream are damaged by this externality (possibly these are fishermen). We could have the same sort of problem with air pollution, too.

Now as we have mentioned previously, the polluting factories do not pay the full cost of pollution. The social cost is the total private cost plus the cost of the pollution to the property owners downstream. Likewise, the full cost of production of a factory that is polluting the air is the total private cost of production plus the lowered values of the other people's property that is damaged by the smoke.

In all the examples, the marginal private cost of polluting is quite small. The marginal social cost is greater because of the resources required to eliminate or reduce such pollution. No one owns the rivers or the air, only the land adjacent to the rivers or under the air. Thus no one sets a price on this scarce resource. In the absence of well-defined property rights, there is no automatic corrective device built into the competitive market mechanism.

What then is the solution to the problem of externalities? While in the next sections of this chapter and in Chapters 7 to 9 we limit our consideration to direct, government-imposed environmental controls, such measures are not the only means by which externalities may be controlled. We conclude our overview with a brief discussion of a few of the other solutions, including environmental controls.

A "solution" that says some sort of external control is necessary gives the impression that the only solution is for government to forbid the polluting firms from discharging waste into the river, smoke into the air, or noise into the ears. However, this type of control may not be socially optimal. The market may well have already "solved the problem." Some property owners choose to purchase property they know is being polluted because they could purchase that property more cheaply than equivalent unpolluted property. Those who suffer windfall losses are owners who acquire property prior to its being polluted and therefore prior to its being made relatively less valuable. The market adjusts the value of the property after it is subject to pollution. We shall return to this issue in Chapter 9.

Alternatively, since we have noted that externalities can arise if property rights are incompletely assigned, government could redefine property rights. In the case of rivers, the river owner(s) might charge the factories for polluting if the downstream landowners own the rivers and in this way make up the loss in property value. Or if the factory is given property rights in the river, the downstream owners could bribe the factories to reduce the amount of pollution. Clearly, in either case there would not be zero pollution. In the case of bribery, if the factory owners owned the river, the downstream landowners would bribe the factories until the marginal cost of an additional bribe equals the value of the marginal reduction in pollution to the downstream landowners. If property rights were assigned to the downstream landowners and a charge levied on the factories per unit of pollution, the factory would pollute until the marginal cost of polluting one more unit equals the marginal return from polluting.[7]

As you have perhaps already recognized, the efficiency of such a solution depends upon the number of parties involved. If 1,000 factories are damaging 10,000 fishermen downstream, it would be very difficult and expensive to work out a transaction; even if an outside party owns the river, the policing costs may outweigh the potential returns. If 9,999 downstream property owners agree to bribe the factories not to pollute and one party does not agree, how could the nonpayer be excluded from the benefits? In the case of one damager versus one damagee, the solution would be simple if property rights were assigned. But the more parties involved, the greater the cost of making the transaction.[8]

Another solution, of course, is for government to force a "solution" by charging the damagers and compensating damagees. Again, there is the question (moral or economic?) of compensating those who acquired property at a lower cost because of the damage. A further question in the case of government control is how much pollution to permit. Surely, a goal of zero air and water pollution is ridiculous. If there is diminishing marginal utility from pollution and increasing marginal costs in terms of resources used in reducing pollution, of course the solution is to have pollution at some optimal, but nonzero, rate. It is frequently the task of economists and engineers to determine that rate. Marginal costs are not easily measured, and in the absence of a social utility function, marginal benefits are generally almost impossible to measure, a point which will be made more clearly in Chapter 9.

A SIMPLIFIED ANALYSIS OF ENVIRONMENTAL CONTROLS

Using the concepts presented, we now turn to a simple analysis of environmental controls. This discussion will provide an introduction to our analysis of the effects of environmental controls on the mineral-extracting industries. Since private property rights to the pollutable resources are not assigned, we will in this section consider only direct governmental solutions to the problem.

Let us begin our analysis in the simplest way with a closed society: "closed" in the sense of having no trade with any other society. This society produces only one good, which the society itself consumes. The only cost of producing this good is the damage that is inflicted upon the environment by production. No labor or capital is used in producing the good. But the greater the rate of production, the greater the rates of pollution and environmental damage.

The people in the society derive utility both from consuming the good produced and from a clean environment. Additional units of the good increase society's utility from goods at the expense of increased environmental damage, resulting in a decrease in society's utility. Society can choose any combination of goods and quality of the environment that is desired. It can pay for a better-quality environment by decreasing the level of output.

In this type of world the only problem is in determining the desired levels of output and environmental quality. Such determination is a simple matter if society consists of only one person or if every person is identical to every other. In either case, society maximizes utility by allowing the production of goods to increase so long as the additional or marginal utility from each increment of goods exceeds the lost utility from the deteriorated environment. Optimization does not permit any additional production of goods for which the marginal utility from the goods is less than the lost utility from the lower-quality environment. In this way society attains the cleanest environment desirable under the specified restrictive assumptions.

Thus far we have implied that the only way society can attain the desired level of clean environment is by a decree setting the level of pollution that is permitted. This situation results because the producers do not bear the cost of pollution and have no incentive to abate. Another method of control does, however, exist if the owner of the producing firm is a profit maximizer and if the society is a monetary society. Once society determines the desired level of pollution, a tax can be imposed upon the production of pollution. The firm continues producing, and hence polluting, so long as the marginal profit from the increased production exceeds the additional tax to be paid on the resulting increased pollution. Once the added tax begins to exceed the marginal profits, the limit to profit-maximizing production is reached. Society of course chooses the tax rate consistent with its level of utility maximization.

We might note in passing that the majority of economists doing research in this area would probably opt for pollution control by taxation rather than by placing controls upon the firms, since taxation would not distort relative prices as much. Governmental policy makers have not been as favorable to taxation as a control method, viewing the tax as a "license to pollute," which is to them undesirable. By this reasoning, similarly, a fine for speeding would be called bad, because it is a license to speed. In any case, since our project here is to estimate the probable impact of controls and not to compare the effect of controls and taxation, we will ignore taxation of pollution for now but mention it later in our empirical analysis.

Thus, we can summarize the major points in our analysis as follows: society can have any quality of environment desired, but a cleaner environment can be purchased only at a cost. In our simplified model the cost is clear and simple. Reduced pollution leads to a reduction in the amount of goods available to the society. As we complicate the analysis to approach more closely the real world, the cost of reduced pollution is not always so readily apparent. It is our purpose to identify these costs, though we do not intend to advocate that any particular level of control be applied. We simply wish to indicate the nature of the economic environment in which society's decision makers are working.

Now let us complicate the model slightly, first by relaxing the assumption that there exists only one person or that everyone is alike. When differences exist among people, there is no clear-cut social-utility function. Each person could independently choose a different level of production and clean environment, depending upon individual tastes. Thus, any level of output chosen results in some consumers' being dissatisfied.

To complicate the issue even further, assume many firms producing many different goods and polluting in widely varying degrees. In addition, assume that the people in the society work for the firms and own the capital stock. Any production restriction now has several effects rather than just one. Restrictions affect the cost of the goods consumed and the state of the environment. They also affect the incomes of those individuals who sell their labor and their capital to the affected firms. There are second-stage effects in cases where other firms deal with the regulated firms. Finally, the regulations affect the level of future investment. It is on this issue of the investment effect that we will concentrate our attention in Chapters 7 and 8.

Thus, the problem of setting an optimal level of production and pollution in a multifirm world is a vastly complicated issue. Certainly, zero pollution is much too expensive even to be considered. But when people differ greatly in how they value goods and the environment, there is little possibility of reaching an optimum situation. Two possible methods of choosing an optimal level can be suggested: (1) voting or (2) allowing market forces to determine the optimal level by redefining property rights. These are examined before turning specifically to an analysis of mineral industries and environmental regulations.

Society can, in theory at least, vote on all possible combinations of production, distribution, and environment. Even if this possibility were physically possible, considerable complications would arise. It is possible or even highly probable that no consistent solution can be reached. To show how possible complications arise, consider the following simple hypothetical example. Suppose there are three individuals, 1, 2, and 3, and three possible events A, B, and C. Each individual has a definite preference ordering when faced with a choice between any two events. The given hypothetical preference ordering is as follows:

1.	(ApB)	(BpC)	(ApC)
2.	(BpC)	(CpA)	(BpA)
3.	(CpA)	(ApB)	(CpB)

In this table (ApB), for example, is read "situation A is preferred to situation B." Note that each individual is rational in the sense that if he prefers A to B and B to C, A is preferred to C.

If this three person society votes between A and B, A gets a majority, two votes to one. Likewise, B would be chosen in a vote between B and C. Rationality would imply that in a vote between A and C, A would win. But it is apparent from the table that in voting between A and C, C is preferred by a majority. Thus, there is inconsistency in determining social welfare, even in this extremely simplified example.[9]

Possibly for this reason, and probably for the reason that voting on each issue is too complex an arrangement in an industrial society, societies vote in other ways. They vote for individuals who specialize in learning about and choosing among alternatives. The motivations of those who actually do the choosing may differ from those of the individuals who are affected, but, apparently, society is willing to bear this cost rather than have each individual learn about and vote on each issue. A major purpose of this study is to provide information to those selected individuals who do make government's policy decisions. We return to these ideas below.

But first let us examine the question of whether market forces can give society the desired level of output and pollution. Market forces can in some cases provide the desired output-pollution combination, and in other cases such forces fail. Let us examine first the case in which consumers, acting through the market, can motivate firms in a particular industry to produce an optimal level of pollution.

If it is assumed that consumers care enough about pollution, the market under some circumstances can motivate firms to abate pollution voluntarily. But the products of the firms doing the polluting must be easily identifiable by consumers to the specific firm. Otherwise consumers are not able to discriminate in their purchases on the basis of a firm's polluting activity.[10]

It appears, therefore, that the market structure under which consumers have the most impact on pollution activity is a product differentiated oligopoly. An oligopoly is an industry with more than one firm, but the number of firms is sufficiently small that all firms' actions are interdependent. Such industries are frequently characterized by a great deal of competition other than price competition, such as advertising, product differentiation, and so on. In some cases, the consumers of the product can be so pollution conscious that they can be marginally motivated to purchase according to a firm's antipollution activities. Thus, for a producer, pollution abatement can be in some sense a substitute for advertising.

Of course, the pollution-abatement activities of the firm increase the price of the product to a greater or lesser extent, depending upon their substitutability with advertising expenditures. Yet it need not follow that product price rises in proportion to pollution expenditure. The pollution-conscious public may have to bear some of the costs of pollution discrimination in the form of paying higher prices for the products of low-polluting firms. Even those firms selling an intermediate product can be motivated to abate pollution at least marginally if those sellers of the final product are motivated to say that they always buy from nonpollutors.

It is not argued that there is empirical evidence to support this analysis. Yet it is stated that theoretically under certain conditions firms may be motivated to some extent by the market to abate pollution. The consumers of the product may have to be sufficiently pollution conscious to bear some of the costs of discrimination in the form of higher prices. And consumers must have information as to which firms pollute and how much each pollutes. In any case, boycotts have in the past worked to the benefit of those doing the boycotting. Certainly, labor-union boycotts are examples. It is noted also that over the past few years firms have advertised their antipollution activities. Such advertising is likely an attempt to increase market shares, capitalizing on an issue having a high public profile. Thus some firms must believe that this type of activity has a positive effect upon sales.

However, in many cases firms may not be motivated by market forces to abate pollution. One case is that of firms that are operating in a competitive market. Competition forces the firm to use the most efficient means of production—that is, the least costly methods in a private sense. In addition, the individual competitive firm sells a homogenous product that cannot be differentiated from the products of other firms by pollution conscious consumers. This results in information costs to the conscientious consumer being high relative to the previous case where producers of differentiated products are boycotted. Finally, the consumers of a firm's product and those affected by the firm's polluting activities must be from the same group. Clearly, some people can be motivated to bear costs in order to cause a decrease in pollution even though the pollution has no effect upon them personally, but it appears that the motivation would be stronger for those actually affected.

It appears from our analysis that most mineral industries do not fit well into the classification of industries most susceptible to market-induced pollution abatement. To be sure, oil-producing firms recently have been advertising their pollution-reducing activities. But, these firms sell gasoline to final consumers in clearly marked stations. Other mineral firms do not generally do this. In fact, as emphasized above, most mineral firms are competitive and sell at a market-determined world price. Thus, the above analysis does not appear to apply. This is not to say that mineral producers may not be charitably motivated to reduce pollution below the profit-maximizing level. Certainly, this motivation can be quite strong. But, competitive forces may not allow for much altruism by firms that must raise capital in a competitive market and sell in a competitive world market in which they compete with many firms that may not be so public spirited. The cost of pollution abatement may simply not be bearable for some individual firms when the market motivation is not present.

SIMPLE-BENEFITS THEORY

The previous analysis of environmental harm can be summarized by noting that the problem arises when purchasers (consumers) are either ignorant of the harm, indifferent to the harm, or unable to afford to eliminate the harm unless all purchasers are forced to contribute. The cost or sacrifice of environmental improvement falls upon individuals: consumers, labor, owners of capital, and so on. Firms do not bear costs *per se*. They are simply institutions through which individuals organize production. Ultimately, individuals benefit or lose from the controls. At the absurd level, even if water control affects only beaver and fish, some individuals receive utility from knowing that some beaver and fish live in cleaner water than they otherwise would have. Let us discuss the methods with which we can calculate benefits.

Since this is an economic analysis, there is justification in concentrating almost exclusively upon economic measures of the benefits of controls and the monetary costs of environmental regulations. After all, this is the only way of comparing the benefits to the cost. It might be asked how the analysis can concentrate solely upon monetary benefits, or how one can measure the value of a person's health or even of his life. Are esthetic values totally ignored? The answer is simple: The researcher does not place values on these factors; the individuals reflect their valuations.

In any case we do not attempt, when estimating pollution damage, to value people's lives or their health; rather, we consider changes in the probability of death or the impairment of health. Of course, no one lives in a riskless situation. Every time someone drives on a busy highway or crosses a busy intersection, there is some probability of death or injury. No one avoids all risks; the cost is too high. People increase their risks because they are not willing to bear the costs of lowering the probability of death.

People volunteered to be kamikaze pilots knowing they were increasing the probability of death. But they knew also that the probability of death was not 100 percent. A much lower percentage was chosen to die, and the reward was high to the remainder who did not die. People were willing to increase the probability of death if they were compensated by an increased probability of reward. This is not a farfetched example. We see construction workers on skyscrapers absorbing a higher probability of death, but they are willing to trade the increased risk for a higher return than they could earn elsewhere. Other examples are test pilots and, of course, miners.

Societies place values on life and health also. There is not a police officer on every corner at which a violent crime may take place; the cost is too high.

For the same reason, society does not put a traffic light on every corner at which an accident could occur.

Environmental conditions are also subject to these considerations. We are not talking about a decision between one state of the environment which is totally pure and another state which kills everyone. A change for the worse in an environment only changes the probability of health damage or even death. People are willing to be compensated for the increased risk. Environmental alterations can produce repugnance or esthetic value, but people can be compensated for the change.

Thus, an individual's valuations of environmental conditions can be expressed monetarily, just as can the valuation of any good. This can be measured by how much he is willing to pay for the good, if he does not own it, or, if he owns the good, by how much he is willing to accept to give up the good. The state of one's environment is a consumption good, just as are the amount of medical care consumed, the amount of food eaten, the number of vacations taken, and so on. No good is any more of a necessity than any other. Hardly anyone consumes any good that he is not willing to give up *some* part of—but not necessarily all—for some additional amount of other goods. This consideration includes environmental conditions.

Finally, the concern may be that some group, possibly a group of poor Indians living near a mine, is simply too poor to assign any monetary value to the damage done to them by pollution. Perhaps this group of Indians survives by fishing, and the water from the mine kills all of the fish in the local stream. Theoretically, we can estimate this cost monetarily. The government can assign property rights in the stream to the Indians, then determine the least amount the Indians would accept for each amount of pollution. This approach gives a perfectly valid measure of damage. Of course, this is not to say that the measurement problem may not be difficult. The technique simply emphasizes that the damage can be estimated monetarily. The problem may actually be that the Indians are poorer than other members of society wish them to be. If so, the problem is not one of measuring damage. It is totally a redistribution problem. The distribution of income, however, has nothing to do with the relevance of measuring the cost of pollution monetarily by willingness to pay.

It appears, therefore, that we are justified in assigning monetary value to the environmental damage. We could then estimate monetarily the net benefits from environmental controls and compare these to the net costs in order to determine, at least theoretically, the optimal level of controls. We are not, however, primarily concerned with determining the optimal level of controls but with measuring the *actual* costs and the *actual* benefits of controls that have already been enacted and with measuring the potential costs and benefits of controls being considered. What is "optimal" is a policy decision. Now we are in a position to develop a more formal theory to examine the effect of environmental regulations and, of course, taxation.

NOTES

1. Most of the discussion in this section is taken from Chapter 11 of C.E. Ferguson and S. Charles Maurice, *Economic Analysis: Theory and Application*, 3d ed. (Homewood, Ill.: Richard D. Irwin, 1978), 498–506. © 1978 by Richard D. Irwin, Inc.

2. Alfred Marshall, *Principles of Economics* (London: Macmillan, 1920); A.C. Pigou, *The Economics of Welfare* (London: Macmillan, 1929).

3. An excellent survey of this literature as it relates to mineral extraction is provided in Gordon L. Bennett, "The Impact of Environmental Controls and Taxes on Mining," unpublished dissertation, Texas A & M University, May 1977.

4. This example is drawn from Otto A. Davis and Andrew B. Whinston, "Economics of Urban Renewal," *Law and Contemporary Problems* (Winter 1961): 105–17.

5. For a more complete discussion of the common-property issue, cf. A. Alchian and H. Demsetz, "The Property Rights Paradigm," *Journal of Economic History* (March 1973): 16–27; R.V. Haveman, "Common Property, Congestion, and Environmental Pollution," *Quarterly Journal of Economics* (May 1973): 278–87.

6. For additional information concerning the issue of property rights, cf. E. G. Furubotn and S. Pejovich (eds.), *The Economics of Property Rights* (Cambridge, Mass.: Ballinger, 1974).

7. In a very famous article by Ronald Coase, "The Problem of Social Cost," *Journal of Law and Economics* (October 1960): 1–44, it was shown that if one party is damaging another through its productive activity, the optimal amount of damage is the same regardless of the party to whom property rights are assigned, given zero transactions costs.

8. This point is particularly well made in J.M. Buchanan and G. Tullock, *The Calculus of Consent* (Ann Arbor: University of Michigan Press, 1965).

9. For further discussion of this issue, cf. Kenneth J. Arrow, *Social Choice and Individual Values*, (New York: Wiley, 1951); Buchanan and Tullock, *op. cit.*

10. For an analysis of the market incentives for the firm to abate pollution, cf. W.P. Gramm, "A Theoretical Note on the Capacity of the Market System to Abate Pollution, *Land Economics* (August 1969): 336–38.

7

Theory of Environmental Controls and Taxation in the Mineral-Extracting Firm and Industry

In Chapter 2 we developed the basic theory of the mineral-extracting firm and industry. Chapter 6 provided a brief introduction to environmental controls and their impact. In this chapter we combine this material with a consideration of taxation in order to analyze the way environmental controls and taxation affect mineral-extracting firms and industries. In our analysis we concentrate on the effect of these exogenously imposed parametric changes on such things as cost, the rate of output, the rate of input usage, and—particularly—investment. As will be shown, the major long-run impact of changes in environmental controls and taxation on the mineral-extracting industries would be expected to come about through changes in investment.

We begin the chapter with brief discussions showing the theoretical reactions of a mineral-extracting firm or industry to both environmental controls and taxation. In these sections we evaluate the short- and long-run impacts, assuming that the firm both buys inputs and sells its output in competitive markets. We show that the result of the imposition of either type

Much of the analysis in this chapter is based on the study performed by G. Anders, W.P. Gramm, and S.C. Maurice, *The Impact of Taxation and Environmental Controls on the Ontario Mining Industry—A Pilot Study*, Ontario Ministry of Natural Resources, 1975. Some of this analysis appears in the study by G. Anders, W.P. Gramm, S.C. Maurice, and C.W. Smithson, *Investment Effects on the Mineral Industry of Tax and Environmental Policy Changes: A Simulation Model*, Ontario Ministry of Natural Resources, July 1978.

of constraint will reduce investment by the firm and industry. Since, however, the theoretical analysis is capable of predicting only the direction of the change, we then turn to the development of a simple analytical model, which is able to predict not only the direction but also the magnitude of changes in investment resulting from changes in environmental controls and/or taxation. Because our model predicts changes in the level of investment using the changes in the rate-of-return to investment (a procedure clearly appropriate in the case of the marginal investment), we follow our discussion of the model itself with an explanation of circumstances in which our methodology would over- or underestimate the effect on *total* investment. Finally, in the appendix to this chapter, we present a brief overview of some of the major articles concerned with the theoretical effects of taxation of mineral-extracting firms and industries. In this review, we concentrate on the more general forms of taxation and defer discussion of the percentage-depletion allowance—a tax treatment specific to the mineral-extracting sector—to Chapter 11.

THE EFFECT OF ENVIRONMENTAL CONTROLS

Environmental controls on mineral-extracting firms can in general take the following five forms:

1. Government limits the amount of some by-product of the firm's output that the firm can put into its external environment per time period.
2. Government limits the total pollutant content that can be present in the firm's external environment (several problems of eternalities are involved in this case).
3. Government may impose emission or ambient controls on a firm's internal or external environment—for example, noise or dust controls on heavy equipment.
4. Government may require that the land surface used by the firm during production be reclaimed in some specified manner at the end of production in that area.
5. Government may require a bond to pay for such reclamation at the beginning of the firm's operations.

With the exception of the ambient and internal controls, to be discussed in Chapter 9, these controls have essentially the same effect on established firms. That effect is to place an additional constraint or additional constraints upon the firm in its attempt to maximize net wealth or present value. As noted above, the constraint can take several forms; but the important point about the added constraint is that the costs, both average and marginal, are increased. This increase affects the firm's output decisions and its use of factors of production.

The marginal and average costs of extraction are increased after a point, because current extraction causes pollution or other environmental damage, and to the extent that the control or constraint is effective, that pollution must be reduced by using costly resources, or the environmental damage must be repaired, again through the use of costly resources. Obviously, until the output is reached at which the constraint is effective, there is no change in costs. But beyond the constraint the full marginal cost is the current marginal cost, plus the marginal user costs, plus the marginal effect upon future costs of extraction, plus the cost of complying with the environmental constraint.

Figure 7.1 indicates the way in which a change in marginal cost affects output with price, P_t, unchanged. The constraint causes the marginal cost of extraction in each period to rise from MC to MC', in turn causing output in period t to fall from q_t to q'_t. This adjustment is, however, strictly a short-run situation.

The firm's average costs are affected also. If some firm cannot earn the going rate of return, this firm will exit in the long run. That is, after imposition of the constraint, the firm will determine whether added investment in order to comply is preferable to closing down. Other firms that are earning above-normal returns will continue operating. If the industry within the constraint-imposing jurisdiction has some effect on the world price, the reduced supply will drive up price. Otherwise the burden of the change will fall on the firms rather than on the buyer of the output. Thus, the effect of the constraint upon output is unambiguous: output will fall. To the extent that the portion of the industry that is affected has an effect upon the price of the output, the price will increase, the increase depending upon the importance of the portion of the industry affected and upon the elasticity of demand for the product.

The effect of the constraint upon the usage of each input or factor of production is neither unambiguous nor straightforward. While the answer could be determined mathematically, this section relies upon literary exposition.

In determining the impact of an environmental constraint or an increase in the severity of the constraint, one must examine how the factors of production are used individually to produce output and either to abate pollution or to restore the environment, depending upon the form of the constraint.

In the absence of *any* constraints, a profit-maximizing, competitive producer hires a factor of production—labor, for instance—until the value of goods (revenues) produced by the last unit of labor hired equals the cost of hiring that last unit. In more formal terms, the value of the marginal product must equal the wage for the firm to be profit maximizing. When the pollution constraint is introduced, the firm adjusts in the same way, taking into

FIGURE 7.1

Output in Period t Where the Environmental Constraint is Included

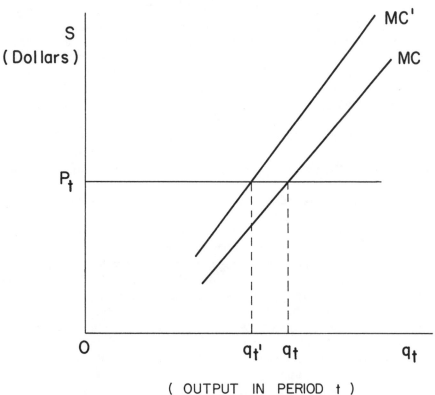

(OUTPUT IN PERIOD t)

consideration all the benefits and costs of hiring each factor. While the explicit cost of hiring the last unit of labor does not change (the firm is assumed to be a price taker in the hiring of factors), there is a fall in the returns to hiring the last unit. The decline in revenue or profits for producing the old level of output can be traced to several factors. Since the last unit of an input increases output, it has a positive marginal product, and therefore the value of the marginal product is positive. But the added product from the last unit hired also contributes to increased pollution, which has, we will assume, a constrained ceiling level. The increased pollution reduces the present value of the firm, or reduces profits to the firm. Therefore, pollution is no longer free to the firm as the present value or profits are reduced somewhat below the level established when unconstrained. The pollution constraint is binding because a factor hired to produce minerals also produces pollution which reduces the firms' revenues or increases the firms' costs. Since revenues are lower than initially while the cost of hiring the last unit of labor is the same, the profit-maximizing firm increases profits (reduces losses) by decreasing the use of these last workers. With the pollution constraint there is a fall in the revenues produced by the workers, so revenues are less than the costs of hiring that factor. No sensible profit-maximizing producer uses a unit of an input that is worth less than it costs.

The same argument can be made for any factor used to produce minerals. It is evident then that moving from an unconstrained situation to a situation in which there is a pollution constraint results in fewer factors hired to produce minerals. An increase in the severity of the constraint has the same effect as enforcing a new constraint; an increase in severity decreases the usage of resources hired to produce output.

But there is now an additional effect upon input usage. The firm that initially faced no pollution constraint now purchases resources for the purpose of reducing pollution or to restore the environment to meet the imposed level. Thus, the amount of labor hired to abate pollution increases from zero to some positive amount. In this case, as is the case of labor used to produce minerals, labor is hired up to the point where the benefits and costs of the last unit hired are equal. The costs are, of course, market determined; but the return to hiring such factors is the reduced pollution, with the result that a firm may now carry mineral production further than the constraint would permit in the absence of abatement. Of course, the concern is with the marginal unit as before, and hiring more such units is not beneficial to the producer when the costs outweigh the value of the returns. Similarly, where equilibrium under a pollution constraint is attained, if the constraint is relaxed somewhat, the marginal unit of labor for abatement has a smaller return but the same cost. Reducing the marginal units thus reduces costs more than revenues, and it is economical for the profit maximizing firm. In summary, when we proceed from no pollution constraint to some constraint, factors to abate pollution are hired, and if the government imposes an additional constraint, the use of factors to abate pollution is increased.

Therefore, when we consider the use of inputs in both occupations, pollution abatement and extraction or production, the effect on employment is indeterminant. We have used labor as our example, but the effect on other inputs is similar.

Let us emphasize that this discussion of the effect upon resource use has glossed over the possibility that some firms may not be able to cover costs in the long run and therefore discontinue production. Certainly, it is apparent that the increased costs from compliance with the pollution constraint could result in marginal firms not earning their opportunity costs and leaving the industry. In fact, if the control or constraint is sufficiently severe, the exiting firms could have previously been earning somewhat more than merely marginal rates of return. Again, this exit would add to the decreased output. To the extent that the domestic portion of the industry cannot pass on much of the higher costs in the form of price increases because it is a price taker in the world market, the effect on output is more severe. But, we should emphasize that because of the two, possibly offsetting, effects upon factor use, we cannot determine unambiguously the total effect of a control on the usage of inputs.

We have until now concentrated on the impact of environmental controls and regulations upon established firms or upon investments that have previously been made. It would appear, however, that the impact upon future investment by either new or established firms will be more crucial and will be our major concern in the remainder of our analysis. In the case of a prior investment the firm would be inclined to ignore fixed cost and try to make the best of a deteriorated situation. It might reduce output or reduce its usage of some inputs but continue to produce at the diminished rate of return, hoping for no further restrictions. Thus, established capital might continue producing even at a rate of return below the rate in the economy as a whole.

This is not the case with new investment. The alternative to a new investment in a mineral industry is investment in other industries, even industries in other countries. In this case the firm would not wish to "let bygones be bygones" and ignore fixed costs. It could consider all alternatives prior to investment. Let us, therefore, before analyzing the effect of controls on investment, review the basic economic theory of investment, already discussed in Chapter 2.

First, let us assume that a firm wants to purchase capital goods (machines) and may do so either by selling bonds or reducing its real money balances.* A stock of machines yields a flow of services in each period; the flow is assumed to bear some fixed relation to the capital stock. The machines wear out eventually, and must be replaced if the flow of services is to be maintained. Since the firm is assumed to be a price taker in the market for loanable funds, it may borrow at the relevant market rate of interest. While the firm may be restricted in its level of borrowing by overall net worth, we will assume that this is not a relevant barrier and ignore such details in the analysis; so for our purposes, the firms may borrow or lend at the market rate.

A profit-maximizing firm desires a stock of capital having a flow of services the present value of which is greater than or equal to the present cost of

---

*For purposes of simplification it is assumed that the supplier of investment funds is external to the firm. This need not be the case. Firms may of course raise funds internally. Such does not change the rate at which funds may be borrowed or loaned. The opportunity cost of funds raised internally remains the market rate of interest.

the capital. This condition is similar to conditions discussed above where we emphasized that revenues from hiring the last unit of labor equal the costs of the last unit. Since a machine yields services for more than one period, the value of services for future periods must be discounted; in other words, one dollar today is worth more than one dollar in a later period. Given a rate of interest r_0, a dollar in any later period t, has a present value of $\left(\dfrac{\$1}{1+r_0}\right)^t$. Of course, the present value of a unit of capital to a firm depends on several factors, such as the market rate of interest, the price of goods produced, the average lifetime of the capital stock, the rate of flow of services from the capital stock, the quantity and price of other factors employed, and so on.

For now, however, we wish to concentrate upon the way in which changes in the market rate of interest affect the investment decisions of the firm. Assume first that a firm faces a given set of investment opportunities in a particular time period and can rank these opportunities according to expected rates of return, from the highest to the lowest. Figure 7.2 shows a hypothetical situation. The marginal or additional rate of return from additional units of investment is plotted on the vertical axis; that amount of investment in the time period is plotted on the horizontal. The downsloping line, the marginal return to investment (MRI), is the locus of all combinations of investment and the marginal rate of return forthcoming from the last unit. We assume for simplicity that prospective capital investments vary continuously.

Assume again that the relevant rate of interest is r_0. The firm invests, during the period, K_0, because at any investment level below K_0 the marginal or additional rate of return from an additional unit of investment exceeds the cost of the unit of investment, r_0; that is, MRI $> r_0$. The firm would not invest beyond K_0, because the rate of return from any added investment would be less than r_0. But if the rate of interest were r_1, rather than r_0, the firm would invest only K_1, during the period. Therefore, the MRI curve or schedule shows the rate of investment forthcoming from each relevant interest rate. This schedule is therefore, the firm's demand for capital during that period. We note that the lower the rate of interest, the larger the rate of investment.

Thus far we have looked at one firm in one industry and have given a simple explanation of the chosen level of capital investment. We can combine all firms in an industry and obtain the industry's demand for capital investment.* The situation is similar; the industry will invest until the level at which the marginal return equals the market rate of interest.

---

*This industry demand is not simply the summation of all firms demands. Any single firm can change output without changing the price of the output. The industry as a whole will affect output prices.

FIGURE 7.2

The Marginal Return to Investment

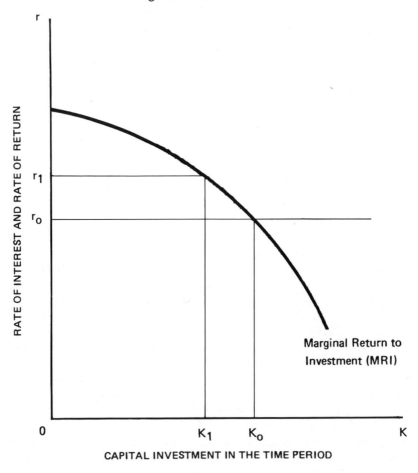

Figure 7.2 The Marginal Return to Investment

But with many industries having profitable projects, the mineral industry is in competition with firms in other industries desiring to produce goods that society demands. If changes in legislation apply equally to all industries, these changes affect every industry. If, however, legislative changes affect some industries and not others, some can be made better off relative to others—i.e., some would choose a smaller share of investment funds than would have been forthcoming in the absence of the rule change.

Suppose initially the laws of the land are well known and all firms are in equilibrium. Let legislation be enacted, making a previously permitted activity illegal. Assume that a pollution ceiling is imposed on all producers but that it is a binding constraint to only a few, including firms in the mineral industry. Now a mineral-producing firm that previously wished to invest K_0 in capital services incurs costs to abate pollution. These costs reduce the present value of the firm's flow of capital services from investment. This reduction results in a downward shift of the marginal return to investment, drawn in Figure 7.3 as a shift from MRI to MRI'. For the firm represented by Figure 7.3 the new level of investment is K_1. The level of investment by the firm and the industry declines as a result of the new constraint. Of course, if the marginal return to investment is driven too far downward, the firm would undertake no new investment. This situation is depicted in Figure 7.3 by MRI'', for which the marginal return from any level of investment is less than the rate of interest.

When fewer projects are undertaken in an industry affected by an environmental constraint, output must fall in the long run. It is more expensive to produce, or alternatively, capital cannot be placed in the affected industry as profitably as before. Of course, the *amount* of the decrease can vary. The effect can be quite small, in which case society might prefer the reduced output because it is offset by reduced pollution. On the other hand, the effect can be quite extensive and damaging to the economy. This estimate of the extent of the investment effect is an empirical question, to which we shall address ourselves in the next chapter.

THE EFFECT OF TAXATION

The effect of taxation upon mineral-extracting firms is similar to the effect of environmental controls. Any real increase in the rate of taxation decreases the after-tax rate of return from a given level of investment or a given capital structure. Certainly, for some firms or industries it may be possible to shift some of the tax forward in the form of higher prices to purchasers of the output. Some may be shifted, too, to factors of production in the form of reduced factor payments. But some part of the increase generally causes a decrease in a country's net returns.

In any case, for industries that sell their input in world markets at world prices, the possibility of a large shift forward is somewhat remote. In addition, for the case of industries that hire rather generalized factors of production (that is, factors not specific to the industry), not much of the tax can be shifted backward to factors of production. Thus, the types of firms and industries with which we are concerned here would generally be affected extensively by changes in the rate of taxation.

FIGURE 7.3

Marginal Return to Investment after Environmental Controls or after Increased Taxation

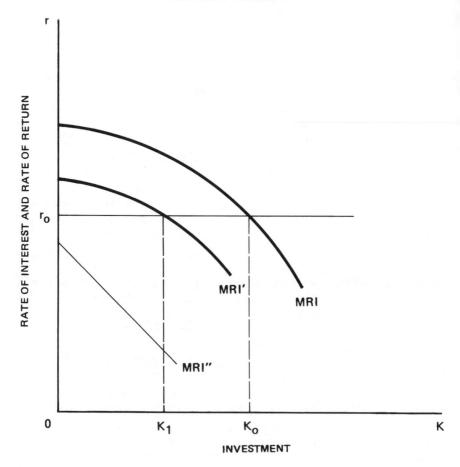

It is quite possible that some firms may be so profitable that the increase in taxation would not drive after-tax returns below the opportunity cost of capital, and the firm would continue to produce. On the other hand, some firms may still continue producing profitably in some areas while phasing out other areas that are not so profitable after the tax increase. Other firms, earning about the going rate of return prior to the tax increase, may find it unprofitable to continue in business over the long run. Again, the amount of decrease in the marginal firms or in marginal areas for more profitable firms depends upon the extent of the tax increase. The amount of shifting is, therefore, an empirical question.

Just as was the case for the imposition of environmental controls, the primary long-run effect of a tax increase is upon the industry's rate of investment. Anything that shifts the marginal return to investment downward must at a given rate of interest cause investment to decline. Since the marginal return must be net of taxation, an increase in the rate of taxation causes such a downward shift.

After the decline in investment because of the increase in taxation, the after-tax return on the marginal investment will in the long run remain constant, other things remaining the same, while the before-tax return will rise. We can show these results graphically with the aid of Figure 7.4. Assume for the sake of simplicity an industry subject to no taxation. With a marginal return to investment of MRI and an interest rate of r_0, the industry invests K_0. Obviously with zero taxation both the before- and after-tax marginal rates of return are r_0. Let some type of taxation be imposed. By our above analysis MRI falls to MRI', reflecting the now lower return from each additional unit of investment after taxation.

With the interest rate remaining constant at r_0, the level of investment falls to K_1. Clearly, the after-tax marginal return remains r_0, but the before-tax return at investment K_1, reflected by MRI, rises to r_1. The extent of the increase in before-tax returns depends of course, upon the extent of the increase in taxation.

Thus far we have simply looked at the direction of change in some variable after an exogenous change in taxation and environmental controls. We turn now to the development of models that will allow us to predict both the direction and the magnitude of the effects within reasonably accurate limits, depending upon the information we possess about certain variables.

SIMPLE PREDICTIVE MODELS

As has been frequently emphasized, competition will in the long run drive the present value of the expected stream of income from an investment to the cost of that investment. As was explained in Chapters 2 and 3 above, when an investment costs less than the expected stream of income, discounted by the appropriate rate, there is the opportunity for returns above the going rate in the economy. Any time there is the opportunity for above-normal returns, investment is attracted into that area. The additional investment has one or more of three possible effects: (1) it can drive down the selling price of the output produced by the capital investment until returns are driven down to the going rate; (2) it can drive up the costs of factors used to produce that output;

FIGURE 7.4

Effect of Taxation on Investment

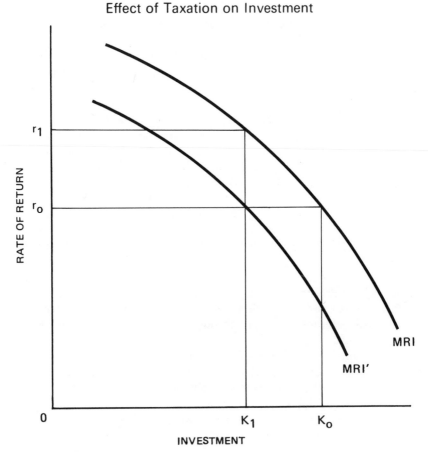

(3) it can result in less profitable investments being undertaken until the discounted return equals the cost of the investment. Similarly, when the cost of investment exceeds the discounted stream of expected returns, investment is withdrawn from that particular area until investments return the going rate in the economy.

We might note that it is the net return *after taxation* that is relevant to the investment decision. It is, after all, the return net of taxation that is available to the investing firm. The before-tax stream of returns will generally exceed the cost of an investment by a substantial amount, but if the expected return after taxation is below the cost of an investment, that investment will not be undertaken.

For analytical purposes, let us assume that two investments yield *identical discounted streams of income before taxation.* The two investments differ because the tax structures and/or the relevant environmental regulations differ. Therefore, the discounted streams of return after taxation and after compliance with the regulations differ. It follows then that if investment under the two sets of conditions would be undertaken until the cost equals the net expected stream of income, the difference in the costs that would be paid for the two identical before-tax streams certainly reflects the extent to which the investment is affected by the different tax structures and environmental regulations.

For example, assume an investment is expected to yield a discounted, before-tax stream of $100. Under one tax structure competition might drive the cost of such an investment to $40, leaving only a normal rate of return. Let the tax structure be changed so that the total tax rate increases. Now investors might be willing to pay only $30 for the investment yielding the before-tax stream of $100, because the after-tax returns are lower. Thus, the ratio of the two costs shows the extent to which investment is reduced from the higher tax rate. In this example, investment to obtain $100 under the new structure would be only 75 percent (30/40) of what would have occurred under the old. Thus, if we can find the extent to which a change in tax rates or environmental regulations changes the amount of investment that would be undertaken to yield a specific stream of income, we can determine the effect of the change on investment.

A Simple Predictive Model for the Impact of Environmental Controls

The imposition of environmental controls, as noted, causes firms to make expenditures they would not have made otherwise. Since these expenditures would not have been otherwise forthcoming, after-tax rates of return on investment must decrease, such decreases causing a diminution of investments in the industry. Almost anyone would agree with this direction of change. The basic disagreement would concern not the direction but the magnitude of the effect.

Many writers and analysts in this area—possibly a majority—argue that the percentage of expenditures required to comply with environmental restrictions is so small relative to total expenditures or total costs that the total effect of controls upon investment and output is probably negligible. Therefore, they say, although the costs to improve the environment may be large in absolute terms, they are so small relative to all other costs that one can ignore their effect.[1] Let us examine these arguments within the framework of a rigorous and formal, but quite simple, economic model. Within this framework we can obtain considerable insight into the total effect upon the rate of investment.

Consider a capital investment expected to yield an income stream for a firm over a 21-year horizon. The expected income stream or present value in time period zero from such an investment can be given by

$$PV = \sum_{t=0}^{20} \left(\frac{1}{1+r}\right)^t (R_t - C_t - F_t) \tag{7.1}$$

where R_t = sales revenue in period t,
             C_t = variable costs in period t,
             F_t = fixed costs in period t (generally these will be front end
                   costs),
             r   = the appropriate discount rate.

Since we are now basically interested in the effect of environmental controls, we will ignore taxes until the next section. Competition will force the cost of such an investment, which we shall call C_I, to the present value of the stream of returns from the investment. For simplicity, let us call the *net return* in period t, Y_t, where

$$Y_t = R_t - C_t - F_t$$

Moreover, call the entire *undiscounted* income stream over the horizon Y, where

$$Y = \sum_{t=0}^{20} Y_t.$$

Call d_1 the *total discount factor*. That is, d_1 is determined such that

$$d_1 Y = \sum_{t=0}^{20} \left(\frac{1}{1+r}\right)^t Y_t.$$

Thus, d_1 is related to r, but it does not equal r.

For example, suppose Y_t, the net income in each year, is $100 and r, the yearly discount rate is 0.10; in this case d_1 is 0.4523, since

$$d_1 = \frac{\$100 \sum_{t=0}^{20} \left(\frac{1}{1+r}\right)^t}{\$2100} = \frac{\$100\,(9.5003)}{\$2100} = 0.4523.$$

Thus, the present value of the assumed income stream from the investment is approximately $950 = 0.4523 \times \$2100$

One would expect firms to undertake such investment yielding Y, so long as the cost of the investment is less than $950. In fact, competition would eliminate excess returns, and one would expect cost to equal present value in the long run.

Next, let us assume exactly the same stream of income, Y_t, with the same discount rate r, but let environmental controls be imposed. As a consequence of such controls, the amount E_t must be spent in some period to comply with the regulation, the value E_t depending upon the period in question. It would be assumed that most of such expenditures would have to come during early time periods. The present value of the stream of returns now becomes

$$PV = \sum_{t=0}^{20} \left(\frac{1}{1+r}\right)^t Y_t - \sum_{t=0}^{20} \left(\frac{1}{1+r}\right)^t E_t = d_1 Y - d_2 E, \qquad (7.2)$$

where $\qquad E = \sum_{t=0}^{20} E_t,$

and d_2 is derived by solving

$$d_2 E = \sum_{t=0}^{20} \left(\frac{1}{1+r}\right)^t E_t.$$

Note that d_2 does not necessarily equal d_1. In fact, as will be shown, it most probably will not.

In the long run, competition will drive the cost of such an investment, C_{II}, toward the present value $(d_1 Y - d_2 E)$. Without the controls, the cost of the identical investment would be C_I. To compare the two situations, take the ratio

$$\frac{C_{II}}{C_I} = \frac{d_1 Y - d_2 E}{d_1 Y} = \left(1 - \frac{d_2}{d_1} \frac{E}{Y}\right). \qquad (7.3)$$

This ratio will then indicate the impact on investment of the environmental controls. For example, suppose that C_{II}/C_I equals 0.90. This means that, after the imposition of the environmental controls, the firms would be willing to pay for a given investment good only 90 percent of the amount they would have been willing to pay prior to the controls; investment would, therefore, decrease by 10 percent.*

---

*This clearly holds for the *marginal* investment (i.e., that investment yielding only the going rate of return in the economy). We will consider the accuracy of such an estimate for *total* investment in the next major section.

Let us now analyze the determinants of the ratio of levels of investment under the two situations. First, it is clear that the effect of controls on investment depends not upon the proportion of required environmental expenditures to total costs but upon the ratio of such expenditures to net profits—certainly a much smaller figure than total costs. Second, the effect of the E/Y ratio upon investment must be weighted by the ratio of the relevant discount factors. There is strong reason to believe that d_2 would generally be considerably greater than d_1. In any case, even if E/Y is rather small, the weighting of d_2/d_1 can strengthen the effect significantly.

The reason that we would expect the environmental cost discount factor, d_2, to be much larger than the income discount factor, d_1, is that the bulk of expenditures to comply with environmental regulation would generally be in the form of front-end costs at the beginning of the period under consideration. On the other hand, income from a mineral investment is often postponed far into the future. If this is the case, (E/Y) is less than (present value of E/present value Y).

To illustrate this effect, consider two income streams of $60 over three years. One is distributed so that there is $5 in year 1, $15 in year 2, and $40 in year 3; the other stream is $40 in 1, $15 in 2, and $5 in 3. Assume a yearly interest rate of 0.10 in both cases. For the first $60 stream biased toward future income, the total discount factor is 0.7834; for the second stream, biased toward the present, the total discount factor is 0.8756. Thus, even under the same interest rate and the same total income the present-biased discount factor is almost 12 percent higher than the future-biased factor.

We gain further insight by extending our analysis of the hypothetical example of the 21-year income stream and the 10-percent yearly discount rate discussed above. Recall that the total income discount factor for the stream of $100 a year for 21 years was 0.4523. Now assume that the same investment under environmental controls requires an additional investment of $100 for abatement in order to comply with the control. It is quite reasonable, too, to assume that the $100 investment must be spent on front-end costs: $50 in year 0, $25 in year 1, and $25 in year 2. Now, the $100 is only 4.76 percent of total net income, which for the average industry would be less than 1 percent of sales. But with the above environmental cost configuration, $d_2 = 0.9343$, a figure more than double d_1. In this case the investment ratio is

$$\frac{C_2}{C_1} = \left(1 - \frac{d_2}{d_1}\frac{E}{Y}\right) = 0.9017$$

This ratio shows that in the above hypothetical example investment to obtain the same stream of income would with controls be only about 90 percent of what would have taken place in the absence of controls, even though the environmental expenditures are only 4.76 percent of total net income and probably less than 1 percent of costs. This example shows the powerful impact of discounting when most environmental expenditures fall at the beginning of the period.

We can use the same type of model to analyze the effect of enforced production delays while the regulations are being enacted and complied with. Assume that output is delayed for three years while the environmental-control expenditures are being undertaken. With the delay the total discount factor for income falls from 0.4522 to 0.3396. The C_2/C_1 ratio consequently falls from 0.9017 to 0.8691. With production delays, the total investment to obtain the hypothesized income stream with controls is only 86.91 percent of what would occur without controls. This example shows the significant impact of delays on investment production.

These hypothetical examples are not meant to be necessarily illustrative of the all mineral extractive industries. They are designed to show that regulated environmental expenditures can have a considerable impact upon the rate of investment in mineral extraction even though they might make up only a small part of total costs. The fact that these added expenses came rather early in the time horizon relative to the net income stream causes them to be heavily weighted in the investment decision.

A Simple Predictive Model for the Impact of Changes in Taxation

The effect on investment of changes in the tax laws can be estimated using the same types of techniques as those described in the example concerning environmental controls. Since there are several types of taxation and methods of depreciating capital equipment, however, the taxation model is somewhat more complicated than the environmental controls model. We will in any case follow the same type of techniques in order to develop a model to estimate the magnitude of the effect of tax changes upon mineral investment.[2]

Let us begin analysis by assuming, as above, that a particular investment is expected to yield a discounted stream of net income *before taxation* of

$$PV = \sum_{t=0}^{H} \left(\frac{1}{1+r}\right)^t Y_t \tag{7.4}$$

where all symbols are as above, and the time horizon of the investment extends from zero to period H. Introducing all relevant forms of taxation that are important for our analysis, we obtain the following after-tax, present-value income from the investment:

$$PV = \sum_{t=0}^{H} \left(\frac{1}{1+r}\right)^t \{Y_t - k(Y_t - \gamma_t C - pY_t - m\rho Y_t) - m\rho Y_t\}, \qquad (7.5)$$

where   C   = the cost of the relevant investment.

                  γ_t = the allowed depreciation of the investment in the t-th period for tax purposes. Thus,

$$\sum_{t=0}^{H} \left(\frac{1}{1+r}\right)^t \gamma_t C$$

is the discounted value of the allowable stream of depreciated tax write-offs.

                 K   = the rate of corporate income tax.

                 p   = the average rate of allowable percentage depletion. Since depletion cannot be applied to all costs, p is not the rate that is used for allowable costs but the average rate for all income.

                 ρ = the percentage of total income from the investment contributed by mining.

                 m   = the average mining tax. Thus, $m\rho Y_t$ is the amount paid in mining taxes. Note that in equation (7.5) mining taxes are deductible from corporate taxes.

For simplicity of exposition and estimation, let the *discounted sum* of all the streams of before-tax income be Y, where

$$Y = \sum_{t=0}^{H} \left(\frac{1}{1+r}\right)^t Y_t,$$

Moreover, let d be the relevant rate of discount associated with depreciation; thus, dC is the discounted value of the stream of depreciated tax write-offs. Again, the term d is not the yearly discount rate or rate of interest. The term d is derived such that

$$dC = \sum_{t=0}^{H} \left(\frac{1}{1+r}\right)^t \gamma_t C_t$$

and therefore reflects both the allowable rate of depreciation, γ_t, and the yearly discount rate, r.

As emphasized above, competition will assure that in the long run investment is carried out until the present value of an investment after taxes equals the cost of the investment. Thus, if in equilibrium an investment costs C, then from equation (7.5)

$$C = Y \frac{[1 - k(1-p) - m\rho(1-k)]}{(1-kd)} \tag{7.6}$$

Equation (7.6) isolates the important features in mineral taxation without complicating the model with types of taxation that are not really relevant to our analysis.

With this basic model, using the fundamentals set forth in the case of environmental controls, we can compare alternative rates of investment under separate tax structures. For example, suppose we wish to determine the amount of investment expenditure that would be used to obtain the same discounted income stream, Y, under different circumstances. Consider two alternative tax structures, one using subscript 0 for the relevant parameter associated with an investment level C_0 for the discounted income stream Y, and the other using subscript 1 associated with investment level C_1 for income stream Y. From equation (7.6) the relative value of the two investments under the two tax structures is given by the ratio

$$\frac{C_0}{C_1} = \frac{[1 - k_0(1-p_0) - m_0\rho_0(1-k_0)]/(1-k_0 d)}{[1 - k_1(1-p_1) - m_1\rho_1(1-k_1)]/(1-k_1 d)} \tag{7.7}$$

One can substitute the best estimate or a range of values to obtain the relative rates of investment under two tax structures. For example, if we wish to isolate the effect of not allowing mining taxes to be deducted from corporate taxes, we would set $k_0 m_0 \rho_0$ equal to zero in the numerator, set $k_1 m_1 \rho_1$ equal to empirical estimates of their values in the denominators, and keep all other estimates the same. The ratio C_0/C_1 would give an estimate of the effect on the amount of investment that would be undertaken to obtain a given stream of income under various assumptions about the estimates of the variables. Or we could let k_0 and k_1 differ to analyze the effect of changes in the corporate tax rate. In fact, we can easily compare the total effect upon investment of two entirely different tax structures (and will do so in the following chapter). Finally, the model is sufficiently general to permit us to introduce any other tax variable we wish to analyze—for example, tax reductions for smelting operations. Clearly, the accuracy of the results depends upon the reliability of the various estimates for the parameter.

The Basic Model

Combining the preceding models, the present value of an investment costing C dollars and having a time horizon extending from period zero to period H can be written as

$$PV = \sum_{t=0}^{H} \left(\frac{1}{1+r}\right)^t \{Y_t - E_t - k[Y_t - E_t - mp(Y_t - E_t) - p(Y_t - E_t) - \gamma_t C] - mp(Y_t - E_t)\} \qquad (7.8)$$

Since we wish to compare the investment expenditures under various assumptions about taxation and environmental regulation, we can calculate the level of investment yielding the same discounted stream of income before taxation and before environmental regulation, under four different assumptions about the cost of the investment:

1. C_I = the cost of an investment yielding a specific discounted stream of income under the basic tax structure and environmental regulations.

2. C_{II} = the cost of an investment yielding the same stream under the basic tax structure but under the altered environmental regulations.

3. C_{III} = the cost of an investment yielding the same stream under the basic environmental regulations and an altered tax structure.

4. C_{IV} = the cost of an investment yielding the same stream under both altered tax structure and altered environmental regulations.

In our prediction models we will use the following symbols:

Y   = the *undiscounted* stream of income from a prospective investment.

d_1 = the total (not yearly) discount rate for the stream of income.

E   = the *undiscounted* stream of environmental expenditures required by the regulations for a particular investment.

d_2 = the total (not yearly) discount rate for environmental expenditures.

C_i = the cost of an investment yielding income stream Y, where $i = I$, II, III, IV. Cost under competitive conditions equals the present value.

d_3^0, d_3^1 = the total discount rate used to depreciate investment costs for tax purposes, respectively before and after tax changes.

k_0 = the corporate income-tax rate prior to the change.

k_1 = the altered corporate income-tax rate.

m_0, m_1 = the average mining-tax rate. 0 is prior to the change; 1 is after the change.

ρ = the average percentage of the income stream subject to the mining tax.

p_0, p_1 = the percentage depletion rates, respectively before and after the tax changes.

Since in the long run the cost of an investment will equal the present value of the income stream, from equation (7.8) the cost of an investment yielding an undiscounted income stream, Y, prior to the tax and environmental changes is

$$C_I = \frac{[1 - (1 - p_0)k_0 - m_0\rho(1 - k_0)]d_1 Y}{(1 - k_0 d_3^0)}.$$

Under the basic tax structure and new environmental regulations,

$$C_{II} = \frac{[1 - (1 - p_0)k_0 - m_0\rho(1 - k_0)]}{(1 - k_0 d_3^0)}(d_1 Y - d_2 E).$$

With the new tax structure and without the new environmental regulations,

$$C_{III} = \frac{[1 - (1 - p_1)k_1 - m_1\rho]d_1 Y}{(1 - k_1 d_3^1)}$$

Finally, with both the new regulations and the new tax structure,

$$C_{IV} = \frac{[1 - (1 - p_1)k_1 - m_1\rho](d_1 Y - d_2 E)}{(1 - k_1 d_3^1)}$$

To examine the effect of the changes upon investment we take three ratios. C_{II}/C_I shows the effect of the change in environmental regulation alone under the old tax structure. For example, if C_{II}/C_I, equals 0.90, we would deduce that the change in environmental regulation alone causes investment to be only 90 percent of the level that would be forthcoming in the absence of such regulation. In other words, the present value of a given undiscounted stream would be 10 percent. C_{III}/C_I shows the effect of the change in the tax structure under the old environmental regulations. Finally, C_{IV}/C_I shows the effect of changing both structures at the same time. Thus, we can either isolate the effect of one or the other of the changes by itself or analyze the effect of both changes. By dividing the relevant costs and simplifying the ratios, it can be shown that

$$\frac{C_{II}}{C_I} = \left(1 - \frac{d_2}{d_1}\frac{E}{Y}\right); \tag{7.9}$$

$$\frac{C_{III}}{C_I} = \frac{[1 - (1-p_1)k_1 - m_1\rho](1 - k_0 d_3^0)}{[1 - (1-p_0)k_0 - m_0\rho(1-k_0)](1 - k_1 d_3^1)}; \tag{7.10}$$

$$\frac{C_{IV}}{C_I} = \left\{\frac{[1 - (1-p_1)k_1 - m_1\rho](1 - k_0 d_3^0)}{[1 - (1-p_0)k_0 - m_0\rho(1-k_0)](1 - k_1 d_3^1)}\right\}\left(1 - \frac{d_2}{d_1}\frac{E}{Y}\right). \tag{7.11}$$

These ratios shown in equation (7.9) through (7.11) are quite similar in structure. The ratio C_{II}/C_I, showing the effect of environmental controls under conditions of no change in the tax structure, turns out to be the same as the ratio derived earlier showing the effect of a change in controls with no taxation. This is not particularly surprising. The total effect of tax and environmental changes, C_{IV}/C_I, is the product of the tax effect, C_{III}/C_I, times the environmental effect, C_{II}/C_I.

Even given a certain amount of intuition, without some estimation of the parameters one cannot state with any degree of reliability the upper bounds of the ratios in (7.10) and (7.11), because of the uncertainty of the tax effect. Clearly the ratios will be positive, but whether they are above or below unity is unclear. The environmental effect, C_{II}/C_I, is clearly between zero and one, and probably much closer to one, since E/Y is thought to be a rather small positive fraction.

Therefore, in order to calculate these ratios one must use estimates of the various parameters: tax parameters, total discount rates, and the ratio of environmental expenditures to net returns. One would probably use several estimates, generally a high, low, and "best" estimate, because for some of the parameters no precise data may be available. This technique will be illustrated in Chapter 8.

COMPARISON OF CHANGES IN PRESENT VALUE WITH CHANGES IN TOTAL INVESTMENT

We have compared and shown how to calculate ratios of the maximum amounts of investment that would be undertaken under various tax structures and environmental regulations to obtain a given stream of income. Certainly, these ratios of present values give the ratio of the amounts that would be invested in marginal investments—that is, investments yielding only the going rate of return on capital in the economy. But, what about investments that yield above-normal rates? Since we are going to use these ratios of present values under different assumptions to make statements about the effect of parametric changes on total investment, we must show that the C_i/C_j ratio does in fact give a rather accurate estimate of the effect on the total amount of investment in an industry. That is, we shall show that if parameters of taxation and environmental regulation change, and if the estimate of the maximum amount that would be invested to obtain any given before-tax discounted stream of income is $C_{IV}/C_I = 0.75$, total investment in the industry under parameter set IV will be approximately 75 percent of what it would have been under set I. It is, after all, the effect on the total investment in which we are interested.

The reader will recall that there are really two general effects of increased investment in an industry that cause the industry's marginal return on investment to decline. First, there is the extreme in which all potential investments are identical, and increased investment increases industry output and hence lowers the price of that output until all investments earn only the going rate of return in the economy. Thus, in any given period total investment equals the total after-tax discounted stream of returns, including opportunity costs, from all investments. At the other extreme, the investors in the industry can rank all potential investments in a period according to profitability. The output of the industry in the particular economy has no effect on the world price of that mineral. Investment is undertaken until the return on the marginal investment equals the going rate of return in the economy as a whole; in other words, the cost of the marginal investment equals its expected present value.

Let us use Figure 7.5 to analyze the first extreme in which all of the investments are the same by assumption. Suppose at first the marginal return on investment schedule for the industry is MRI_1 in panel B. If the relevant rate of interest is OR, the total investment during the period is OI_1. The return on each investment is driven down to OR, because the price of the output is driven down with increased investment. Next, in panel A, we plot total investment along the horizontal axis and total after-tax present value from all investments along the vertical axis. The unit of measure along both axes is dollars.

The 45-degree line from the origin shows that for every level of investment, the cost of the investment equals, by definition, the amount of the investment. The curve PV_1 shows the total present value of all investments for any level of investment. At low levels of investment total present value exceeds the total cost of the investment, because output price does not fall sufficiently to eliminate all above-normal returns. At the equilibrium level, OI_1, there are no above-normal returns to any investment, and total present value equals total investment. Of course, beyond OI_1, total present value would be less than total investment.

FIGURE 7.5

Comparison of Investment Changes

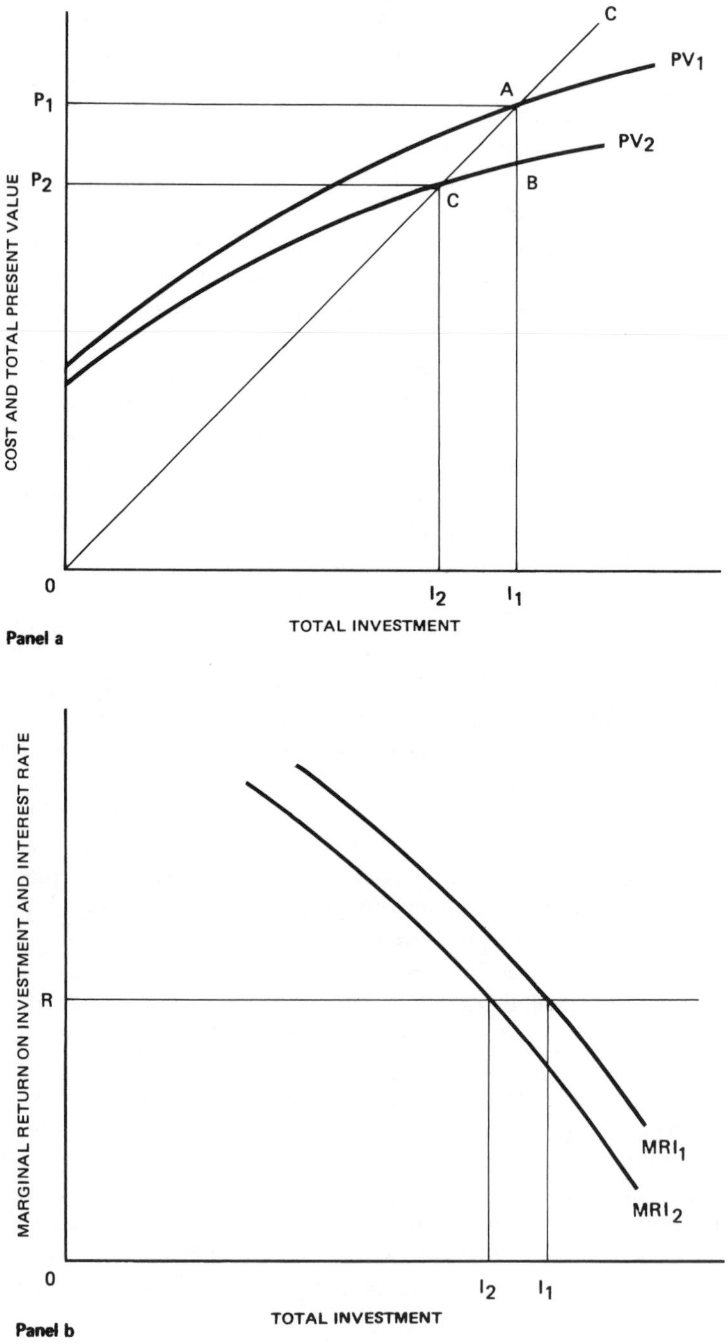

Panel a

Panel b

164

Suppose now that some parameters change and, in the manner discussed in the preceding section, decrease the present value (or the maximum amount that would be paid) of any investment by k percent. If the return from any investment falls by k percent, the total return from all investments (total present value) at any level of investment falls by k percent also. This effect is shown in panel A as the shift from PV_1 to PV_2. (Note that the distance between PV_1 and PV_2 increases as investment increases, because the distance represents k percent of a larger value.)

Under the original set of parameters, equilibrium is at A with investment level OI_1. With the new set of parameters equilibrium is at C, representing investment level OI_2. If above-normal returns on all investments are eliminated, the marginal return on investment must fall to MRI_2 since the interest rate presumably remains the same. The distance AB represents a fall of k percent in present value. But investment must decrease by more than k percent. That is, if AB is k percent of OP_1, $P_1 P_2$ must be more than k percent of OP_1; thus, $I_1 I_2$, the decline in investment, must be more than k percent.

No matter what the shape of the original PV curve, total investment falls by a greater percentage than the percentage fall in the present value of any investment, unless, of course, the total present value curve is a horizontal line, in which case the percentage decline in present value equals the percentage decline in total investment. Thus under the very restrictive assumptions set forth here, the ratio of C_i to C_j, the effect on marginal investment, is an underestimate of the effect of parameter changes on total investment.

Let us turn now to the other extreme in which potential investments differ in expected rates of return, and industry output has no effect on world price. For intramarginal investments that yield a return in excess of the going rate in the economy, the cost of the investment will be less than the expected after-tax present value. This is not to say that under competition investors will not drive the cost of the investment up to the expected present value. This may well be the case, particularly for mineral investment. Suppose many investors believe that a particular property, whether leased or held in fee simply, will yield a given present value. One would expect the price of the property to be bid up to its present value. But there will be no change in real mineral investment because of the bidding up of the price of the property lease. Cost to the investor will equal present value, but the higher cost simply represents a transfer from the investor to the original property holder. The analysis is not changed. Some investments yield a return in excess of the going rate of interest. In these cases the present value is greater than the real cost; i.e., present value divided by cost exceeds one. In such cases environmental restrictions and increases in the rate of taxation would reduce the after-tax present values of investments but only at the expense of rents. Nonetheless, total investment would decline. Let us address ourselves to the problem of how much total investment falls for a given percentage decline in expected present value.

Under the extreme case now under consideration, suppose that for any expected, before-tax, discounted stream of income, $C_{IV}/C_I = 0.75$. Thus, the after-tax present value of any before-tax present value expected from an investment is 25 percent lower. Obviously, less investment would be undertaken under the new tax and environmental laws, but will total investment in the industry decrease by 25 percent also, or will it decline by more or less than 25 percent? The answer clearly depends upon the way in which the marginal return on investment curve shifts after the structural changes.

To approach the problem most easily, consider Figure 7.6. Total investment during a period is plotted along the horizontal axis. The ratio of present value to the discounted stream of the cost of the marginal investment is plotted along the vertical. The curve I_1 shows the ratio for each marginal investment. Since investments can be ranked from highest yield to lowest, the ratio declines as total investment in a period increases. As we have shown, investments in an industry will be undertaken so long as the expected after-tax present value exceeds costs. Thus, if I_1 shows the ratio of after-tax present value to cost, investment will take place until this ratio equals one. As shown by the graph, total investment in this case is OT_1.

Next, let a structural change cause the expected, after-tax present value of each investment to fall by 25 percent. The cost remains the same. Thus, the ratio for each investment falls by 25 percent. This decline is shown by the 25-percent fall of I_1 to I_2. Clearly, investors will invest until the ratio equals unity, and investment will fall.

In Figure 7.6, the shift from I_1 to I_2 is drawn so that the distance represented by CB is a 25-percent decline, and the decrease in investment from OT_1 to OT_2, represented by AB, is a 25-percent decrease. But this need not be the case; the distance AB could represent more or less than a 25-percent decrease, depending upon the position of the curve. Just because the ratio curve shifts downward by a certain percentage, it need not shift leftward by the same percentage.

The question is an empirical one. We can approximate a measurement by returning to the familiar MRI curves. In Figure 7.7, MRI_1 is the original after-tax marginal return on investment curve for a particular industry. This curve shows the declining rate of return for each additional investment as total investment increases. We will assume that it declines for both of the extreme reasons: The price of the output falls as investment increases, and potential investments can be ranked according to expected rates of return, from highest to lowest.

If the relevant rate of interest is OR, total investment in the period is OI_1. Let the tax and environmental laws change so that MRI falls from MRI_1 to MRI_2. This represents a 25-percent fall in the rate of return on each investment. As shown, total investment falls from OI_1 to OI_2, or the amount AC. As the figure is drawn the reduction in investment is in fact 25 percent, but,

FIGURE 7.6

Relation of Percent Changes in Present Value to Percent Changes in
Total Investment

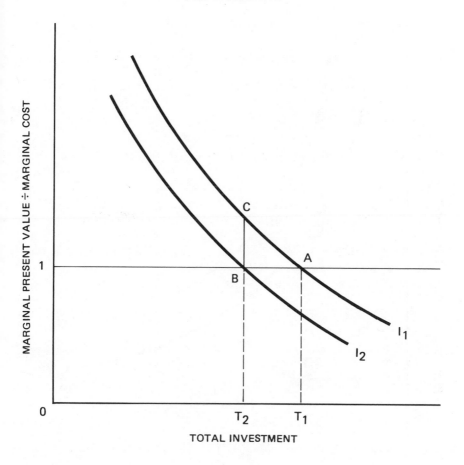

depending on the position of the curves and the way in which they shift, this
need not be the case.

There is no way to estimate empirically how the curves shift. For small
shifts we can assume that if total investment has unitary interest elasticity, the
percent of downward shift will approximately equal the percent of leftward
shift. Unitary elasticity means that along an MRI curve the percent decline in
the rate of return on investment equals the percent increase in total investment.

FIGURE 7.7

Comparison of Changes in Total Investment with Percent Changes in the Present Value of Investment

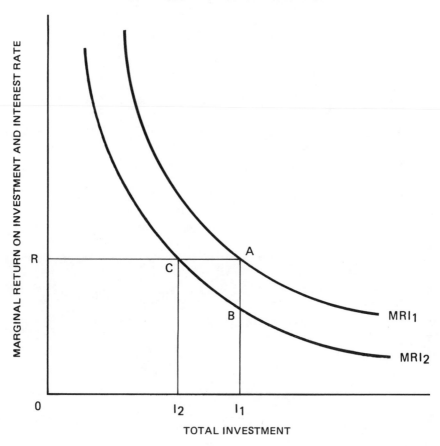

In terms of MRI_2 the percent decline represented by the distance AB along MRI_2 equals the percent increase represented by the distance CA.*

To the extent that the MRI declines because of price changes, our estimate of the effect of changes in taxation and environmental controls on total investment will be an underestimate. To the extent that MRI declines because of decreasing returns to investments, our estimates will be accurate (depending of course on the accuracy of the parameters used to make the estimate) if the elasticity of MRI is reasonably close to one.

CONCLUSION

In this chapter we have developed a basic methodology that can be used to evaluate the impact of environmental controls and changes in taxation. In very general terms, the procedure allows us to estimate the magnitude of the effect on investment in the mineral-extracting industry by comparing the rates of return from an investment before and after the imposition of environmental controls or a change in taxation. In Chapter 8, we will adapt this methodology and actually perform some estimations using data drawn from Canadian mining in order to show how the model can be used.

We might in passing emphasize that changes in environmental laws and in the tax structure have two basic effects: (1) They affect the output rate and even the survival of established investments, and (2) they affect the rate of investment in the future. Our analysis seems to show that the effect on investment is probably more important, in that this has greater long-run repercussions. But, we might note before concluding this chapter that the possibility of future changes in the institutional framework can and will change the future profitability of investment made now and may have a depressing effect upon investment when there is considerable uncertainty about the way in which these changes might occur. Investors greatly apprehensive about future tax increases or about tightened regulations will certainly require a much higher rate of return for an investment to be undertaken than they would require otherwise. Therefore, the effect of changes

---

*In the course of this research, the authors performed some preliminary estimations to ascertain the validity of this assumption of unitary interest elasticity of investment demand. These estimates, reported in Anders, et al., *op. cit.*, indicated that the data from the extractive industries (i.e., mining and oil and natural-gas extraction) are consistent with our assumption of unitary elasticity. More specifically, our estimates showed that, at a 95 percent confidence level, unitary interest elasticity could not be rejected, although a perfectly inelastic investment demand (i.e., elasticity equal to zero) can be rejected.

on established firms may well influence future investment. This effect is not as suitable to empirical analysis, however, as those analyzed above.

NOTES

1. An example of this type of argument may be found in the study sponsored by the American Council on Environmental Quality and the U.S. Environmental Agency (Council on Environmental Quality, *The Economic Impact of Pollution Control: A Summary of Recent Studies* [Washington, D.C.: Government Printing Office, 1972]). This study contains no analytical framework that could be employed to evaluate the impact of environmental controls; it simply assumes that the costs of controls are small and will be absorbed out of operating income. For a review of the study cf. Gordon L. Bennett, "The Impact of Environmental Controls and Taxes on Mining," unpublished dissertation, Texas A&M University, May 1977.

2. This subsection and the above section dealing with environmental controls follow to some extent a method of analysis developed by Arnold C. Harberger in "The Taxation of Mineral Industries," *Federal Tax Policy for Economic Growth and Stability* (Washington, D.C.: Joint Committee on the Economic Report, 1955), pp. 439–49. The basic points were developed further in Harberger, "The Tax Treatment of Oil Exploration," *Proceedings of the Second Energy Institute* (Washington, D.C.: The American University, 1961), pp. 256–69. These papers are reprinted as Chapters 11 and 12 of Arnold C. Harberger, *Taxation and Welfare* (Boston: Little, Brown, 1974). See Chapter 11 for a discussion of these papers.

APPENDIX:

A SURVEY OF
THE LITERATURE CONCERNED WITH TAXATION OF
MINERAL-EXTRACTING FIRMS AND INDUSTRIES

There has been a considerable body of literature concerning the effects of various types of taxation of mineral extracting firms and industries. The first such economic analysis was in a paper by Lewis C. Gray[1] (the theoretical section of which we summarized in the appendix to Chapter 2). The analysis of taxation follows directly from his theory of mining. Assuming no exploration costs for the discovery of additional deposits, Gray pointed out that a tax on the product of a mine can effect the supply from that mine, depending on the type of tax. For instance, an annual tax on the value of the mine causes the present rate of extraction to increase at the expense of future extraction and increases the amount of ore marketed. However, a tax on the yearly surplus of a mine does not change the rate of extraction, according to Gray. Any alteration that causes a loss in surplus must reduce that loss less than the tax saving. If, however, the tax is seen as temporary, extraction is pushed toward the future. This type of tax has no effect, therefore, if and only if the tax is recognized as permanent. A tonnage tax of a fixed amount per ton causes a slower rate of extraction. That is, since future returns are discounted, the tonnage tax reduces the relative returns of present to future extraction, causing a bias toward the future. Gray concluded that even a tax that takes the entire rent and some of the royalty will not affect supply if the relation between present and future returns is not altered. Again, it must be noted that Gray ignored replacement cost in his analysis.

Hotelling in the paper mentioned previously addressed himself to some of the same taxation problems that Gray analyzed.[2] He discussed Gray's point that an unanticipated tax on the value of a mine has no effect upon the rate of extraction. He also proved mathematically that an anticipated tax of a specific rate on the value of a mine has the same effect on the rate of extraction and the value of a mine as an increase in the rate of interest of exactly the tax rate. He proved mathematically that a severance tax per unit of ore extracted tends to bias extraction toward the future. For an exhaustible supply this tax on a monopoly eventually lowers the market price of the resource under Hotelling's limiting assumptions. The length of the period of extraction is extended. In

171

periods in which price is higher because of a severence tax the incidence of the tax falls heavily upon the monopolist. Hotelling did, however, point out that the total wealth of the community might fall from such a tax. We might again stress that his analysis is based upon a fixed, known, and exhaustible deposit.

In 1967 Mason Gaffney edited a volume that contained several papers (two of which were summarized in the appendix to Chapter 2) that were given at a symposium on the taxation of extractive resources. We will summarize here parts of two of these papers. The first was by John D. Hogan, who wished to analyze the effects of various tax policies upon exploration and extraction under less restrictive assumptions than those usually made.[3] He suggested that such an analysis is far too complex for anything other than a computer-simulation model and developed the requirements for such a model. However, he specified only the form of such a model without drawing implications, so no real conclusions can be drawn.

In the same volume Henry Steel discussed the effects of different forms of taxation upon mineral-extracting firms.[4] His analysis differed from most prior analyses because of his emphasis that firms must carry out exploration in order to reproduce their exhaustible resources. In his discussion of how mineral firms should be taxed, Steel pointed out that there can be no purely objective criteria for deriving conclusions on simple equity grounds. One can, however, draw conclusions based upon efficiency grounds.

Steel began by assuming that all mineral rights are owned by the state. The state can do two things: (1) auction off mineral rights competitively, or (2) capture all rents with lump-sum taxes. He stressed that only under extremely limiting assumptions—perfect knowledge; perfect competition; tastes, technology, and resources given; among others—can welfare be maximized by extracting all rents, and either method of approach mentioned would give the same results. But let us emphasize these assumptions are very restrictive and could never be realized in the real world.

Steel then described the effects and problems involved in relaxing some of the assumptions he first specified. He pointed out that the rents involved in mineral extraction are mostly in the form of quasi rents, which involve a return to capital. These rents are reduced as the resource is depleted. Moreover, under risk, some funds must be involved in exploration to replace the depleted resource. In any case there has been considerable controversy concerning the way in which replacement costs—exploration—should affect the method of taxing the returns to mineral extraction. Steel had an extensive discussion of this problem, which we will not summarize here. He did reach the conclusion that *only* in cases in which a deposit is inexhaustible would the rent doctrine— that the state can tax all rent with no effect—apply. When some royalties are necessary for the replacement of capital, the doctrine must be modified. The amount of rent that can be taxed without effect depends upon the need for more exploration. Risk, uncertainty, and ignorance in the real world strengthen Steel's conclusion.

After Steel examined those peculiar characteristics that contribute to uncertainties in mineral-extracting industries, he turned to the effects of several types of taxes upon these industries, under the assumption of positive exploration and replacement costs along with uncertainty. The first tax is a property tax. Theoretically, as noted above, a property tax can be used to change the rate of extraction over time. Steel pointed out that this method of altering production is probably a rather slow process in reality. The rate of change in extraction from a change in the tax rate depends upon the extent to which cost increases from increased exploitation. Furthermore, if the firm is a monopoly, the total amount of change depends upon the extent to which prices change with output. But, to the extent that the resource is monopolistically controlled, which would imply a smaller rate of output than the competitive rate, current output can be increased and current prices lowered with a property tax. The effect, according to Steel, would be quite minor except under rather unusual circumstances. The monopolist also produces less at a higher price in future years with the tax than without.

Steel then distinguished between specific and ad valorem severance taxes. Clearly, a specific severance tax per unit of ore causes a switching from extraction of lower-grade ores to extraction of higher-grade ores, because relative marginal costs change. Such taxes lower output and increase prices to the extent that the taxes are shifted. An ad valorem tax on revenue would not discriminate against lower-grade ores. The burden of the tax varies with product prices. Specific severance taxes decrease output in depressed periods and increase it during prosperity. Ad valorem taxes reduce output and increase prices.

Any severance tax discriminates against the taxed mineral, of course. It also discriminates in favor of more profitable mines, because the tax does not vary with the cost structure. If the tax did vary with costs, it would retard extraction of high-grade, low-cost ores. Another possibility of a severance tax suggested by Steel involves an excessively rapid exploitation of a valuable resource by a few exploiters. Because of intense competition, price is low and output high. A severance tax would increase price and reduce present output. But, with a relatively inelastic demand, the tax would have to be quite high. At any rate Steel suggested that the analysis assumes that the governmental authority somehow "knows" the "proper" rate of exploitation and incidentally the form and rate of technological advance. He pointed out that much of the conservation movement may merely keep in the ground resources for which future generations have little or no use. He might also have noted that conservation could also involve one generation's giving up resources, and hence income, to make a future generation, which will probably be richer than this generation, even richer. These points will be analysed in detail in Chapter 10.

Steel then argued that severance taxes could be used to stimulate or regulate mineral extraction on public lands. He did, however, indicate that a better alternative might be public competitive auction of mineral rights, which would enable the public to capture the quasi rents. But, the uncertainties of exploration would drive the bidding price away from the rate that merely extracts the rents. The problem is, who bears the risk? Certain institutions do develop, such as the sharing of a payment for mineral rights between a bonus and a royalty. Steel indicated that the leasor really bears the risk. Unless the leasor paid for the mineral rights, his sole risk is that he may not extract all that he could under certainty; this is still a risk, however, even though it does not involve an explicit cost outlay.

Steel's section on severence taxes concludes with several ethical arguments along with the suggestion that if government wishes to accelerate the development of certain publicly held minerals, it can impose a low royalty rate and low ad valorem severance tax. Another conclusion is that ad valorem rather than specific severence taxes are probably preferred on grounds of both conservation and economic stability.

Steel next analyzed the effect of business income taxes under uncertainty. He first noted the considerable disagreement among knowledgeable economists about the effect of a corporate tax. Some say that little of the tax can be shifted to consumers, causing a great shift in investment from the corporate to the noncorporate sector; others say that possibly 100 percent of the tax is shifted, in which case consumers bear the tax burden (discrimination remains because prices are increased in the corporate sector). In either case as Steel pointed out, there is little basis for the tax on equity grounds.

Even if the more profitable firms are taxed relative to the less profitable, efficiency may well decline. Not only might it be the case that high profits reflect efficiency; high profits might also reflect underinvestment in the industry. There are also political criticisms of the corporate income tax based upon the fact that they frequently reduce exports, particularly when the firms are export-oriented mineral firms.

One important criticism of the corporate income tax in the case of mineral extraction described by Steel was originally set forth by Stephen L. McDonald in an earlier paper, which we will discuss at greater length in Chapter 11.[5] This paper caused considerable controversy at the time of publication. McDonald argued that a fully shifted corporate income tax discriminates against the riskier and more capital-intensive industries relative to the less risky and less capital-intensive. The riskier industries experience a greater price increase because their tax base is a larger percentage of their price. The change in relative prices causes a reallocation away from the riskier industries. McDonald also argued that mineral industries are on average, but not exclusively, riskier and more capital intensive than manufacturing. As we shall show, McDonald used this argument as a justification for a percentage depletion allowance for minerals, varying in degree from industry to industry.

The final tax considered by Steel is a value-added tax. There are two types of value-added taxes: (1) the income types, which tax total revenue from sales, less depreciation and less purchases (but not purchases of depreciable assets), and (2) the consumption types, which tax sales revenue less purchases, including purchases of depreciable assets, but not depreciation. The impact of these taxes on price, production, and exploration depends upon the extent to which the taxes are shifted. Relative product prices and relative factor prices must remain constant. Steel showed that value-added taxes discriminate less against risky industries than do corporate income taxes, even though the former are not completely neutral in their effect on production. But he also demonstrated with some examples that a value-added tax even with shifting will considerably reduce the range of relative price changes as compared with a corporate income tax. Thus, the value-added tax may well be more neutral than other types of taxation. It is unclear as to the discriminatory effects of a value-added tax on industries with different levels of risk.

The summaries of these few papers should give some idea as to the general thinking on the theory of taxation of natural resources. The primary concern is the effect of specific types of taxes upon the future stream of extraction and upon the extent of investment among industries. The analysis basically attempts to forecast whether a particular type of tax biases extraction toward the present and away from the future, or vice versa. Some analyses also attempt to predict whether a tax is neutral—i.e., whether it affects relative prices and resource uses between natural resource industries and other types of industries. Economists are probably in basic agreement concerning the effects of most types of taxation. We might note, however, that they are not in agreement about how much of a tax can be shifted. Moreover, as will be described in Chapter 11, there is a certain amount of disagreement among economists about the fundamental effect of a percentage depletion allowance for natural resources.

NOTES

1. Lewis C. Gray, "Rent under the Assumption of Exhaustibility," *Quarterly Journal of Economics* (May 1914): 464–89. Reprinted in Mason Gaffney (ed.), *Extractive Resources and Taxation* (Madison: University of Wisconsin Press, 1967), 423–46.

2. Harold Hotelling, "The Economics of Exhaustible Resources," *Journal of Political Economy* (April 1931): 137–75.

3. John D. Hogan, "Resources Exploitation and Optimum Tax Policies: A Control Model Approach," in Mason Gaffney (ed.), *op. cit.*, 91–106.

4. Henry Steel, "Natural Resource Taxation: Resource Allocation and Distributional Implications," in *ibid.*, 233–68.

5. Stephen L. McDonald, "Percentage Depletion and the Allocation of Resources: The Case of Oil and Gas," *National Tax Journal* (December 1961): 323–36.

8

An Estimation of
the Effects on Investment in
Mineral-Extracting Industries of
Changes in Tax and
Environmental Legislation:
The Case of Canada

In this chapter our purpose is to provide an illustration of the manner in which the effects of tax and environmental policy changes may be empirically examined. More precisely, we are interested here in the effect of these structural changes on investment in the mineral-extracting industries. In order to accomplish this task, we will use the analytical model developed in Chapter 7 to evaluate changes in taxation and environmental policies that occurred in Canada in the early 1970s. Specifically, the policy changes to be considered are (1) the corporate income tax rate was reduced; (2) the percentage depletion allowance was changed to earned depletion only; (3) the provincial mining tax rates were increased; (4) mining taxes were no longer deductible from corporate incomes taxes; and (5) environmental controls were tightened substantially. We use these institutional changes in the Canadian mineral industry to illustrate the application of the model developed in Chapter 7 because so many important changes occurred over a very short period of time. Frequently, such institutional changes occur gradually, but the Canadian mineral-extraction industry experienced a significant structural change in that brief period. Thus we can use several before-and-after values of variables in

Much of the analysis contained in this chapter first appeared in G. Anders, W.P. Gramm, S.C. Maurice, and C.W. Smithson, *Investment Effects on the Mineral Industry of Tax and Environmental Policy Changes: A Simulation Model* (Ontario Ministry of Natural Resources, July 1978).

our simulation equations to show how these equations can be used by policy makers to predict the results of structural changes.

We begin by presenting the estimates of the parameters to be employed in our model. We then present our actual estimates of the impact of these changes on the mineral-extracting industries in Canada. Finally, we present some evidence that corroborates our estimates.

ESTIMATES OF PARAMETERS

Obviously, no one could expect to estimate precisely and accurately every relevant parameter in our predictive equations. Such terms as the average effective rates of percentage depletion allowances or the relevant rates of discount for each expense item are not known with any great precision. We, therefore, use ranges of estimation for most terms. After consultation with experts from government and industry, we came up with "best" estimates for each parameter. With expert advice, we were able to obtain a high and a low estimate for the terms, too. In this way we are able to make our estimates of the effects of certain changes in the parameters within a certain range.

Table 8.1 shows the estimates of the tax-related parameters we use. Table 8.2 shows the total discount rates used. Our estimates for the required environmental expenditures as a proportion of *before-tax income* net of costs after the changes in environmental legislation are, for the case of new mines where the new environmental constraints can be engineered into the original design,

$$E/Y = \begin{cases} 0.04 & \text{High} \\ 0.03 & \text{Best} \\ 0.02 & \text{Low} \end{cases}$$

TABLE 8.1

Terms under the Prior System of Taxation and under the New Tax Structure

	Prior System				New Tax Structure			
	Term	High	Best	Low	Term	High	Best	Low
Corporate income-tax rate	k_0	—	0.5	—	k_1	—	0.46	—
Percentage depletion	p_0	0.333	0.25	0.18	p_1	0.25	0.15	0
Percent of total income subject to mining tax	ρ	1.00	0.75	0.60	ρ	1.00	0.75	0.60
Mining-tax rate	m_0	0.195	0.133	0.085	m_1	0.275	0.202	0.06

TABLE 8.2

Total Discount Rates

	Net Income d_1	Environmental Expenditures d_2	Investment Depreciation before Change d_3^0	Investment Depreciation after Change d_3^1
High	0.55	0.88	0.73	0.78
Best	0.40	0.76	0.62	0.67
Low	0.30	0.65	0.50	0.55

The nonstatutory parameters here and elsewhere in this chapter were estimated by policy analysts of the Ontario Ministry of Natural Resources familiar with the relevant statutes and public company financial data and prior to use here submitted to major mining companies for evaluation. This provided assurance that the estimates were "in the ball park" without difficulties over disclosure of confidential data.

In the case of existing mines, new constraints would require massive redesign and modification—a much more costly process. For this case, only one estimate was used: $E/Y = 0.15$

In Table 8.1 the federal income tax does not have a high or a low estimate; we use the statutory 50 percent before the change and 46 percent after. The estimates of the percentage depletion allowance are lower after the change in tax structure, reflecting the shift to earned depletion. On the other hand, the high and best estimates of the provincial mining taxes rose, while the low estimates fell. The provincial mining-tax rate for Canada is a weighted average of the estimates for the individual provinces.

The estimated total discount rates in Table 8.2 reflect the bias of the relevant income or expense stream either toward the present or toward the future. The discount term for net income is the lowest because income from an investment has the longest of the three time horizons. The discount rate for environmental expenditures, is much higher, because the bulk of such expenses are front end costs, coming as they usually do early in the time horizon. Finally, the total discount rate for the depreciation of an investment, d_3^0 (prior to the tax changes) and d_3^1 (after the tax changes) fall between the other two. The rate after the tax change, d_3^1, is slightly higher than d_3^0, reflecting the more rapid allowable depreciation under the new tax structure.

Before using the values of the parameters to obtain estimates about the effect of parametric changes in investment rates, we checked to see whether our estimated figures were consistent with expert estimations of the *effective* (as opposed to statutory) federal income-tax rate. The effective rate is the actual rate applicable after deductions. The estimates we obtained for the effective rates of taxation before and after the changes in the tax structure are

	Before Change	After Change
High	0.29	0.35
Best	0.25	0.29
Low	0.22	0.26

Next, we use our "best" estimates of the relevant parameters and discount rates prior to the changes in order to obtain our "best" estimated effective tax rate. We will use in these estimates the mining tax rate for the province of Ontario. Since the actual income stream chosen as a base does not affect the ratios, we will assume a nondiscounted, before-tax income stream of Y = \$100. The equation showing the long-run cost of an investment yielding \$100 prior to the parameter changes (C_1), using "best" estimates, yields

$$
\begin{aligned}
C_1 &= \frac{[1-(1-p_0)k_0 - m_0\rho(1-k_0)]\,d_1\,Y}{(1-k_0 d_3^0)} \\
&= \frac{[1-(1-0.25)0.5-0.08\times 0.75(1-0.5)]0.4\times 10}{(1-0.5\times 0.62)} \\
&= \frac{0.595}{0.69}\,40 = \$34.49
\end{aligned}
\tag{8.1}
$$

The total effective tax on the \$100 stream of income, using the cost given by (8.1), is

$$
\begin{aligned}
ET &= k_0[(1-p_0-m_0\rho)Y - d_3^0 C_1] \\
&= 0.5[(1-0.25-0.08\times 0.75)100 - 0.62\times 34.49] \\
&= \$23.81.
\end{aligned}
\tag{8.2}
$$

Since the estimated tax rate in (8.2) is derived from the parameter estimates, we can check the consistency of the estimates supplied by government and industry experts by comparing the estimates of equation (8.2) with the estimates of effective tax rates. Using estimates for effective rates and distributions, the estimated tax is 23.81 percent. This is a figure quite close to the "best" estimate of the tax experts (25 percent).

We now carry out the same operation using the $100 income stream, the investment cost C_{IV}, and the best estimates of parameters after the structural changes in taxation and environmental regulations. Under these assumptions, the cost of an investment yielding $Y = \$100$ is

$$C_{IV} = \frac{[1-(1-p_1)k_1 - m_1\rho](d_1 Y - d_2 E)}{(1 - k_1 d_3^i)} \tag{8.3}$$

$$= \frac{[1-(1-0.15)(0.46)-(0.15)(0.75)][(0.4)(100)-(0.76)(3)]}{1-(0.67)(0.46)}$$

$$= \$27.09$$

The total effective tax payment is

$$ET = k_1[(Y-E)(1-p_1)-d_3^2 C_{IV}] = 0.46[(97)(1-0.15)$$
$$-0.67(27.09)] = \$29.58,$$

which is, of course, 29.58 percent of $100. This figure is extremely close to the "best" estimate of the rate (29 percent) and suggests the consistency of the estimates employed.

IMPACT ON INVESTMENT

This section combines the basic model developed in Chapter 7 with the various estimates of the parameters in Section 8.2 to estimate the effect on investment of the changes in the tax structure and in environmental regulations. To this end we estimate C_{II}/C_I, C_{III}/C_I, and C_{IV}/C_I, using all relevant values of the important parameters. The ratio C_{II}/C_I shows the effect on investment due strictly to changes in environmental regulation alone, assuming the tax structure is left unchanged. The ratio will show the extent to which investment would have been undertaken under the changed assumptions, relative to what would have occurred under the original assumptions. The second ratio indicates the extent to which investment would be diminished, assuming no change in environmental regulations but all of the important changes in the tax structure. The final ratio gives the effect of the combination of these two effects.

We used all relevant values of the ratio of discount rates for environmental expenditures and income, d_2/d_1, using the three estimates of each of these discount rates set forth in the preceding section. We combined these ratios with each of the estimates of E/Y, the proportion of the net stream of returns required to be spent on investment conforming to environmental requirements. The results are presented in Table 8.3.

The estimates of the extent to which investment would be undertaken under the regulations, relative to that forthcoming in their absence, is given in the last column of Table 8.3. As can be seen, the effect varies between a 2-percent and an 11-percent decrease in mineral investment in Case A,* using constraints on new investments, and between 18-percent and 44-percent in Case B, using constraints on old investments. The conclusions derived using all of our "best" estimates of the relevant parameters is that under the new environmental regulations, assuming the old tax laws, mineral investment would have been approximately 95 percent of what would have occurred in the absence of the environmental regulations for Case A, and 72 percent for Case B. This, of course, means that, if required expenditure to comply with environmental regulations is 3 percent of net income, mineral investment would decline 5 percent in Case A. For Case B, if the required expenditures represent 15 percent of net income, investment would decline by 28 percent. This emphasizes our prior statement that the impact of environmental expenditure is multiplied because of the difference in the rates used to discount the two streams, income and expenditure. The figures in the body of Table 8.3 show each of the computed reduction in investment and the borders show the values of the parameters used to carry out the estimates with which these impacts were obtained. The same method can also be used to obtain estimates of the effect of an even greater increase in the severity of the controls.[†]

We turn next to the much more complicated estimate of the impact of changes in the structure of taxation upon the rate of mineral investment under the assumption that environmental regulations remain the same. The estimates are much more complicated because of the far greater number of parameters that can affect the range of estimation. The equation for the ratio C_{III}/C_I, which predicts the effect of the change in tax laws, has nine distinct parameters: corporate-tax rates before and after the change, k_0 and k_1; percentage depletion allowances before and after, p_0 and p_1; before and after provincial mining tax rates, m_0 and m_1; before and after discount rates for investment depreciation d_3^0 and d_3^1; and the average percentage of income for which the mining tax applies, ρ. As was shown in Tables 8.1 and 8.2, except for the federal income-tax rates, each parameter has three separate estimated

---

*Because we are primarily interested in a range of estimates, we have rounded to two decimal places. For example, the last figure is 97.64 rather than 98 percent.

[†]In carrying out our computations we set the rates of discount on investment, d_3^0 and d_3^1 equal in order to reflect our assumption of no changes in the taxation structure.

TABLE 8.3

Relevant Values of (d_2/d_1), (E/Y), and (C_{II}/C_I)

(Environmental Impact without Tax Change)

d_2/d_1	A: New Constraints Applied to New Projects (E/Y)				B: New Constraints Applied to Existing Mineral Producers.		
	0.04	0.03	0.02	(C_{II}/C_I)	(E/Y)	$(d_2/d_1)(E/Y)$	(C_{II}/C_I)
2.93	0.11	0.08	0.05	0.89	0.15	0.4395	0.56—low
2.53	0.10	0.07	0.05	0.90	0.15	0.3795	0.62
2.20	0.08	0.06	0.04	0.92	0.15	0.3300	0.67
2.16	0.08	0.06	0.04	0.93	0.15	0.3240	0.68
1.90	0.07	0.05*	0.03	0.94	0.15	0.2850	0.72—best
1.62	0.06	0.04	0.03	0.95*	0.15	0.2430	0.76
1.60	0.06	0.04	0.03	0.96	0.15	0.2400	0.76
1.38	0.05	0.04	0.02	0.97	0.15	0.2070	0.79
1.18	0.04	0.03	0.02	0.98	0.15	0.1770	0.82—high

*Indicates Value Based on "Best" Approximation of d_1, d_2, and (E/Y).

values. Thus, there are 2,187 possible combinations of values, and hence 2,187 possible estimates of the ratio.*

Therefore, for ease of exposition some simplifying assumptions are necessary. First, we assume, probably rather realistically, that for the depletion allowance, the mining tax, and the investment depreciation discount rate, if the high estimate is relevant before the change, the high is correct after the change; similarly "best" and low estimates are always used together. If the high p_0 is used to estimate a particular ratio, the best p_1 or the low p_1 will not be used in the same computation. Different levels of estimation of the same parameter are never combined. Furthermore, we do not use as estimated values when calculating our tables the low figure for postchange provincial mining tax rates. We feel that an average rate of 0.06 is far too low to give meaningful results. For our estimates of the effect for Canada we do not use the lowest rate of depletion allowance, 0.00, when combined with both the highest mining tax rate and the highest discount rate for investment depreciation. We feel that this combination yields estimates of the effect of changes in taxation that are far too high.

Following these assumptions, Table 8.4 shows the estimated impact upon investment of the change in taxation for each combination of parametric estimates used. Column 1 shows the percentage of investment forthcoming under the new tax structure relative to that forthcoming under the old; that is, the extent to which investment would be reduced under the new structure is given by 1.00 minus the figure given in column 1. Columns 2a through 5b show the parametric values associated with the impact figure in column 1. Obviously, because of rounding, several sets of parametric values may be associated with the same impact figure.

The range of impact lies between 0.86 and 0.63, showing between a 14-to 37-percent reduction in mineral investment. The Figure associated with the "best" estimates of the parameters is 0.79, or a 21-percent decline in investment. The starred figures in the table show the estimated parameters associated with the "best" estimate. The average impact is 0.741, or a 26-percent decline.† Thus, the reduction in investment is probably somewhere

---

*With three possible values of seven parameters the number of possible combinations is three to the seventh power.

†The standard deviation is 0.066; so, with the mean of 0.79, the range of the estimate is 0.772 to 0.808, using a 95-percent level of confidence. Let us emphasize that the relevance of our range of estimation depends upon whether we assume our parameters are random variables drawn from a population of all relevant values of the parameters. If these are considered the only parameters, then the mean is simply what it is, and there is no point in estimating a range. Finally, if we used all of the parametric values assumed to be totally unrealistic, for example, a mining-tax rate of 0.06, the high impact figures associated with the omitted values yield an average of 0.86 or a 14-percent reduction in investment. This figure is not far from the average we computed under our assumptions.

TABLE 8.4

Estimates of the Impact of Taxation Changes upon Mineral Investment as a whole (C_{III}/C_I)

Impact C_{III}/C_I (1)	Depletion Rates Before p_0 (2a)	Depletion Rates After p_1 (2b)	Percent of Income Subject to Mining Tax ρ (3)	Mining-Tax Rates Before m_0 (4a)	Mining-Tax Rates After m_1 (4b)	Discount Rates for Depreciation Before d_3^0 (5a)	Discount Rates for Depreciation After d_3^1 (5b)
0.86	0.333	0.250	0.600	0.133	0.202	0.500	0.550
0.85	0.333	0.250	0.600	0.133	0.202	0.620	0.670
0.84	0.333	0.250	0.600	0.133	0.202	0.730	0.780
	0.250	0.150	0.600	0.133	0.202	0.500	0.550
0.83	0.250	0.150	0.600	0.133	0.202	0.730	0.780
	0.250	0.150	0.600	0.133	0.202	0.620	0.670
0.82	0.333	0.250	0.750	0.133	0.202	0.500	0.550
0.81	0.333	0.250	0.750	0.133	0.202	0.730	0.780
	0.333	0.250	0.750	0.133	0.202	0.620	0.670
	0.333	0.250	0.600	0.175	0.275	0.500	0.550

Associated Parameters

0.80	0.333	0.250	0.600	0.195	0.275	0.730	0.780	
	0.333	0.250	0.600	0.175	0.275	0.620	0.670	
	0.250	0.150	0.750	0.133	0.202	0.500	0.550	
0.79*	0.250	0.150	0.750	0.133	0.202	0.730	0.780	
	0.250*	0.150*	0.750*	0.133*	0.202*	0.620*	0.670*	
	0.250	0.150	0.600	0.195	0.275	0.500	0.550	
0.78	0.250	0.150	0.600	0.195	0.275	0.730	0.780	
	0.250	0.150	0.600	0.195	0.275	0.260	0.670	
0.76	0.333	0.250	1.000	0.133	0.202	0.500	0.550	
	0.180	0.000	0.600	0.133	0.202	0.500	0.550	
	0.333	0.750	0.750	0.195	0.275	0.500	0.550	
	0.180	0.000	0.600	0.133	0.202	0.620	0.620	
0.75	0.333	0.250	1.000	0.133	0.202	0.130	0.780	
	0.333	0.250	0.100	0.133	0.202	0.620	0.670	
	0.333	0.250	0.750	0.195	0.275	0.730	0.780	
	0.333	0.250	0.750	0.195	0.275	0.620	0.670	
	0.180	0.000	0.600	0.133	0.202	0.730	0.780	
0.73	0.250	0.150	1.000	0.133	0.202	0.620	0.670	
	0.250	0.150	1.000	0.133	0.202	0.500	0.550	
	0.250	0.150	0.750	0.195	0.275	0.620	0.670	
	0.250	0.150	0.750	0.195	0.275	0.500	0.550	
0.72	0.250	0.150	1.000	0.133	0.202	0.730	0.780	
	0.250	0.150	0.750	0.195	0.275	0.730	0.780	
	0.180	0.000	0.750	0.133	0.202	0.620	0.670	
	0.180	0.000	0.750	0.133	0.202	0.500	0.550	

TABLE 8.4 (Continued)

	Associated Parameters							
Impact	Depletion Rates		Percent of Income Subject to Mining Tax	Mining-Tax Rates		Discount Rates for Depreciation	
C_{III}/C_1 (1)	Before p_0 (2a)	After p_1 (2b)	ρ (3)	Before m_0 (4a)	After m_1 (4b)	Before d_3^0 (5a)	After d_3^1 (5b)
0.71	0.180	0.000	0.750	0.133	0.202	0.730	0.780
	0.180	0.000	0.600	0.195	0.275	0.500	0.550
0.70	0.180	0.000	0.600	0.195	0.275	0.730	0.780
	0.180	0.000	0.600	0.195	0.275	0.620	0.670
0.67	0.333	0.250	1.000	0.195	0.275	0.620	0.670
	0.333	0.250	1.000	0.195	0.275	0.500	0.550
0.66	0.333	0.250	1.000	0.195	0.275	0.730	0.780
0.65	0.180	0.000	1.000	0.133	0.202	0.500	0.550
	0.180	0.000	0.750	0.195	0.275	0.500	0.550
0.64	0.250	0.150	1.000	0.195	0.275	0.500	0.550
	0.180	0.000	1.000	0.133	0.202	0.730	0.780
	0.180	0.000	0.750	0.195	0.275	0.620	0.670
	0.180	0.000	1.000	0.133	0.202	0.620	0.670
	0.180	0.000	0.750	0.150	0.275	0.730	0.780
0.63	0.250	0.150	0.100	0.195	0.275	0.730	0.780
	0.250	0.150	0.100	0.195	0.275	0.620	0.670

*Shows "best" estimate.

TABLE 8.5

Total Impact upon Investment of Changes in Taxation and in Environmental Regulations ($C_{IV} C_I$)

Values of C_{III}/C_I (Taxation)	ENVIRONMENTAL CASE "A" Values of C_{II}/C_I (Environmental)				ENVIRONMENTAL CASE "B" Values of C_{II}/C_I (Environmental)		
	0.98	0.95*	0.92	0.89	0.823	0.715*	0.561
0.86	0.843	0.817	0.791	0.740	0.708	0.615	0.483
0.83	0.813	0.789	0.764	0.739	0.683	0.593	0.466
0.79*	0.774	0.751*	0.727	0.703	0.650	0.565*	0.443
0.76	0.745	0.722	0.699	0.676	0.625	0.534	0.426
0.73	0.715	0.694	0.672	0.650	0.601	0.522	0.410
0.70	0.686	0.665	0.644	0.623	0.576	0.501	0.392
0.64	0.627	0.608	0.589	0.570	0.527	0.458	0.359

*"Best" Estimates of Parameters.

between 20-percent and 26-percent. It is certainly between a 14- and a 35-percent decline, a quite substantial impact.

Table 8.5 shows some *total* impact estimates. These total estimates are derived using the ratio C_{IV}/C_I, which combines both the tax impact and the environmental impact. We present only a few of the values in order to give some feel for the general range of impact. Anyone interested in other predictions can easily carry out the calculations for other figures that they consider more realistic. Using the "best" estimates of the impact we get a 25-percent reduction in investment for environmental Case A, and a 43-percent reduction in Case B.* Other values can be obtained from the tables. The reader will note the tremendous difference in the impact according to whether the new environmental constraints are applied to existing mines.

---

*Given the importance of the province of Ontario in Canadian mining and the fact that Ontario's mining-tax rate is lower than the national average, we also estimated the impacts shown on Tables 8.4 and 8.5 for Ontario. The results were extremely similar to those for Canada as a whole. These results are reported in Anders et al., *op. cit.*

Before ending this section we might stress that our predictions of reduced investment are not meant to indicate the reduction below the level that investment *has been in the past*. They are our estimates of the reduction below the level that *would have occurred* in the absence of the regulation. That is, if, for example, 0.789, our best estimate, is in fact correct, then over the long run mineral investment will be approximately 21-percent lower than what would have occurred under the old taxation and environmental legislation. Furthermore, other changes in such types of legislation may affect the prediction also.

SOME SUPPORTING EVIDENCE FROM BEFORE-TAX AND AFTER-TAX RATES OF RETURN

The period after the recent institutional changes in the structure of mineral industries has not been sufficient in length for a test of the empirical validity of our estimations. For this reason we cannot show a substantial amount of statistical evidence about our predictions. Furthermore, the period prior to 1970–71 witnessed so little change in the tax structure that regressions over this earlier period would almost necessarily show very little relation between rates of taxation and mineral investment. That is, the effect upon investment of changes in such variables as wage rates and mineral prices, both of which fluctuated much more than tax rates, would tend to dominate the effect of tax rates, which changed hardly at all.[1]

We can, however, gain considerable insight into the effect of the change in the institutional structure by examining the difference between the rates of return on capital in mining before and after taxation. This additional evidence can be compared with our predictions about the effect of the changes, in order to check the consistency of our estimates. In this discussion we group both the changes in environmental controls and changes in taxation under the single heading of "taxation." As we have shown in the preceding section, the effect of environmental controls on the firm's rate of return is the same as that for taxation. Indeed, environmental controls may well be viewed as tax collected in kind: The firms are required to make expenditures they would not otherwise make, rather than pay money into the federal treasury.

The reader will recall that it is the after-tax rate of return that is relevant to the investment decision. In the long run, investment in an industry will be carried out until the prevailing after-tax rate for the society as a whole is obtained. Thus, if society's rate of return on capital is relatively unchanged, an increase in the tax rate for one segment of the economy must be followed in the long run by an increase in the before-tax rate of return in that segment.

This effect is shown graphically in Figure 8.1. The rate of return on *additional* investment in an industry is plotted along the vertical axis; the

FIGURE 8.1

Effect of a Change in Tax Structure

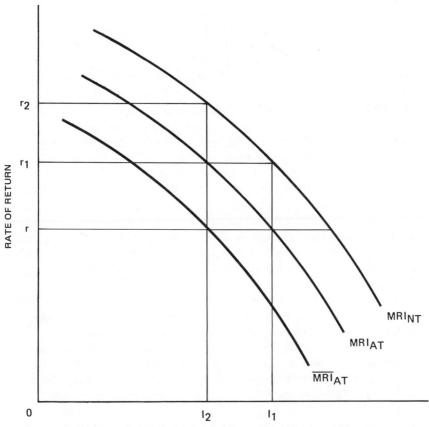

MARGINAL RETURN ON INVESTMENT BEFORE AND AFTER TAXATION

amount of investment is plotted on the horizontal. The reader will recall that the MRI is the return from additional units of investment. The MRI schedule is downward sloping to reflect the assumption that the more profitable investments are undertaken first; as new investments are added, the returns to additional or marginal investments are reduced. In Figure 8.1 MRI_{NT} is the schedule of marginal returns on investment before any taxation. Obviously, the MRI schedule net of taxation, here MRI_{AT}, lies below MRI_{NT}. The greater the effective rate of taxation, the further the after-tax schedule lies below the before-tax schedule.

Let the relevant interest rate or social rate of return be "r". Since investment is undertaken as long as the after-tax return from added investments exceeds the going rate of return, a rate of "r" leads to an investment level of I_1. From the graph the before-tax return associated with an investment level I_1 is r_1, given by the MRI_{NT} schedule. Thus, the difference in the before- and after-tax rates of return $(r_1 - r)$ reflects to a large extent the effect of taxation. Next, assume that the rate of taxation increases. Holding the before-tax schedule constant, the increase in taxation drives down the after-tax MRI schedule, in our example to $\overline{MRI}_{AT}$ in Figure 8.1.

If the rate of return in the total economy remains unchanged at "r," and there is no reason that it will not do so in our example, the rate of investment falls to I_2, because the net returns are now reduced. The difference in returns $(r_2 - r)$ now reflects the effect of taxation. Thus, the *change in the difference* between before- and after-tax rates of return $(r_2 - r_1)$ shows in large part the effect on investment of the change in taxation in the long run.*

A relatively constant difference between before- and after-tax returns in an industry over a period of time would be some evidence of little change in the tax structure over that period. A substantial increase in the difference between rates would indicate a significant increase in the effective rate of taxation. If after-tax returns remained about equal to the rate of return for the economy, it would indicate that the reason for the increased difference in before- and after-tax returns is a decline in investment. That is, the less profitable investments are not undertaken, leaving only the more profitable investments.

Obviously, in the short run some of the increase in the difference between before- and after-tax returns would possibly show up as a fall in after-tax returns on previously constructed investments. However, as argued above, this deviation would be eliminated in the long run when differences would reflect decreased investment and higher before-tax returns. Thus, we would predict that an increase in effective tax rates in one segment of the economy would at first be reflected in after-tax returns somewhat lower than the going rate for the economy. Later (the length of the period depending upon the adaptability of the factors of production), the after-tax rate rises as unprofitable investments are no longer used. The before-tax rate of return will increase in response to the decrease in the rate of investment.

---

*Obviously, the extent of the difference depends also upon the steepness of the slope of the MRI curves.

Let us now analyze actual rates of return in the mining industry to examine the effect of recent tax changes and to see whether the changes had the predicted effect. We use for rates of return, as we did in Chapter 3, the profits-to-capital-employed ratio and the profits-to-equity ratio.[2] These ratios are generally thought to be the best estimators of rates of return.

As shown in Table 8.6, the difference in rates of return before and after taxation were extremely stable for both rates of return—profits to capital and profits to equity—for both total mining and metal mining from 1962 through 1969, and even through 1971 in the case of total mining. This stability of the differences is more clearly seen in Figure 8.2 for total mining and Figure 8.3 for metal mining. As these graphs indicate, the differences were extremely stable in the early years of the period. Thus, investment must have been little affected by the tax structure.

TABLE 8.6

Difference in Before-Tax and After-Tax Rates of Return in Canadian Mining Industry

	Total Mining		Metal Mining	
Year (1)	Profit-to-Equity Ratio (2)	Profits-to-Capital-Employed Ratio (3)	Profit-to-Equity Ratio (4)	Profits-to-Capital-Employed Ratio (5)
1962	0.019	0.015	0.023	0.019
1963	0.018	0.016	0.023	0.018
1964	0.023	0.019	0.030	0.026
1965	0.024	0.019	0.034	0.028
1966	0.022	0.019	0.028	0.024
1967	0.022	0.018	0.029	0.025
1968	0.025	0.021	0.034	0.029
1969	0.024	0.020	0.031	0.025
1970	0.032	0.026	0.055	0.043
1971	0.014	0.011	0.017	0.013
1972	0.032	0.024	0.042	0.030
1973	0.051	0.038	0.072	0.050
1974	0.107	0.080	0.121	0.085
1975	0.097	0.070	0.067	0.047

Source: Statistics Canada

FIGURE 8.2

Difference in Before-Tax and After-Tax Rates of Return, Total Mining

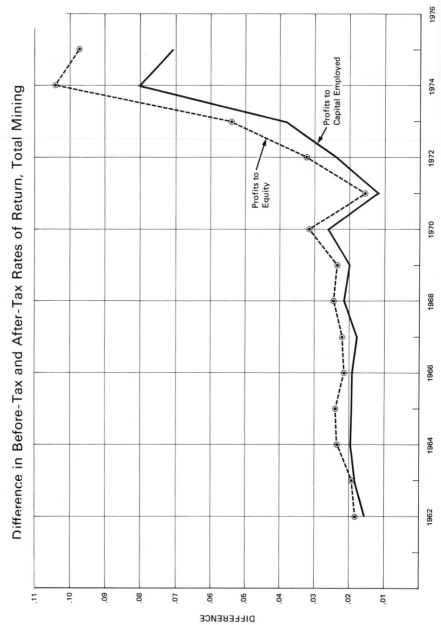

FIGURE 8.3

Difference in Before-Tax and After-Tax Rates of Return, Metal Mining

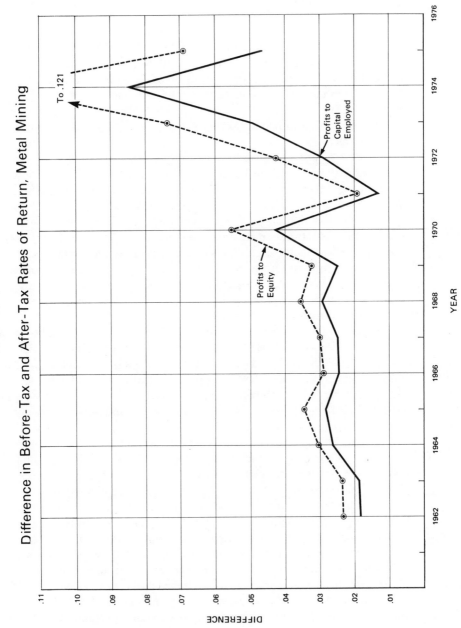

Note how dramatically the differences in before- and after-tax rates of return increased during the 1970s. This substantial increase indicates a much more severe structure of taxation and a consequent decline in investment. The pattern is shown in all columns of Table 8.6 for both metal and total mining. The pattern is much more clearly seen in Figures 8.2 and 8.3. Note how greatly and how rapidly the differences increased after 1971 in all four cases examined. In fact, in all cases the differences trace smooth paths during the early years of the period. During the later years the differences increase dramatically and fluctuate considerably. This increase in the difference shows that the tax burden increased substantially in mining. The severe fluctuations probably indicate in addition a great deal of uncertainty during this period concerning future tax changes.

One might protest, however, that the increase in the difference between before- and after-tax returns between the 1960s and 1970s simply reflected a general economy-wide trend and, therefore, does not indicate a significant change relative to the economy as a whole. One could argue that the rather severe rate of inflation during the 1970s caused the differences in before- and after-tax rates of return to increase throughout the economy. That is, if output and input prices increased at the same rate because of inflation, after-tax returns and before-tax returns could increase at the same rate, even though tax rates remained constant. Since the difference between rates would increase at the same rate, that difference in rates of return would increase if the capital value does not increase at the inflationary rate.

But the evidence is not consistent with this point of view. In the first place, if inflation had been the basic cause of increases in the differences between returns, all differences should have risen together, but this did not occur. Using data from the sources described in Chapter 3, we calculated the increase in the *average* difference between before- and after-tax returns between the periods 1962–69 and 1970–75 for mining, metal mining, and total manufacturing; the results are shown in Table 8.7. While the average difference in the profit-to-equity ratio for manufacturing increased 25 percent from the 1960s to the 1970s, the difference in the same ratio increased 117 percent for metal mining and 155 percent for total mining. Similarly, the difference in the profits-to-capital ratio increased 17 percent in manufacturing, compared to an 80-percent increase in metal mining and a 122-percent increase in all mining. This shows a significant change in the impact of taxation on investment and is, therefore, counter to the above-stated argument against our evidence.

There is further, and perhaps more convincing, evidence against the argument that inflation rather than changes in the tax structure, which diminished investment, caused the significant increase in the difference between before- and after-tax rates of return in mining. This evidence is that the *percentage* of difference increased substantially during the 1970s. The reasoning is as follows. Suppose that inflation and the resulting effect on input

TABLE 8.7

Percentage Increase in Differences between Before-Tax and After-Tax Rates of Return from 1960s to 1970s

	Profit to Equity	Profit to Capital Employed
Mining	155	122
Metal Mining	117	80
Manufacturing	25	17

and output caused both revenues and costs to increase by, say, 10 percent. Then before-tax profits would also increase by 10 percent, profits being defined as revenue minus cost. If the total capital and the tax rate remained relatively constant, then the difference in the rate of return on capital would increase by 10 percent. This can be easily seen algebraically:

Let B = before-tax profits.
             t = the rate of taxation, $0 < t < 1$.

Thus, $(1 - t)B$ = after-tax profits.

Let C = the amount of capital.

In the absence of inflation the difference in the rates of return to capital before and after taxes would be

$$\frac{B}{C} - \frac{(1-t)B}{C} = (t)\frac{B}{C}.$$

The *percentage* difference in the rates, defined as the difference in rates divided by the after-tax rates, would be

$$\frac{t(B/C)}{(1-t)(B/C)} = \frac{t}{(1-t)}.$$

As in our example, let input and output prices increase by 10 percent, causing before-tax profits to increase by 10 percent. Now the difference in rates is

$$\frac{(1.10)B}{C} - (1-t)\frac{(1.10)B}{C} = \frac{(1.10t)B}{C}$$

In this case, 10-percent inflation causes the difference to rise by 10 percent. In the case of percentage differences with 10-percent inflation, the result is

$$\frac{(1.10t)B}{C} \div \frac{1.10(1-t)B}{C} = \left(\frac{t}{1-t}\right).$$

This is the same ratio as that derived before the 10-percent inflation. Thus, inflation alone does not cause the percentage difference to increase. For the percentage difference to increase, the before-tax return must increase relative to the after-tax return. This result is what we expected; and the evidence is consistent.

Table 8.8 shows the percentage differences for both ratios in metal mining and in total mining from 1962 through 1975. Note again the stability prior to 1970 in all four cases. Note also the substantial increase in the 1970s, an increase consistent with our estimates for the absolute differences. This shows the identical pattern described above. The percentage of difference becomes much larger and much less stable during the 1970s. Figures 8.4 and 8.5 show much more vividly the trend we have described for mining and metal mining. This is strong evidence for our argument that the rate of effective taxation increased significantly and caused a substantial weakening of investment.

Now we might summarize what these results mean in the mineral-extracting industry. First, clearly taxation became more severe. This result is consistent with our estimates. Second, the difference between before- and after-tax returns could have come about both by a decrease in after-tax rates of return and an increase in before-tax returns. As argued in Chapters 2 and 3, a decrease in after-tax returns below the going return in the economy as a whole cannot and does not persist in the long run. Therefore, in the long run an increase in the effective level or rate of taxation will result in less profitable investments not being undertaken or being abandoned, which will occasion an increase in before-tax rates of return, the strength of the increase depending largely upon the degree of increase in rates of taxation.

Obviously, the period 1970–75 is too short for all long-run results to take place. One would expect some part of the impact of increases in taxation to come about through a decrease in after-tax returns, the decrease in investment and concomitant increases in before-tax returns occurring over a longer period. But, as shown here, there is some indication that the predictions or estimates derived in the previous section are already beginning to come about. The trend is particularly noticeable in total mining. The average after-tax profit to equity ratio rose 0.001 from the 1960s to the 1970s, while the equivalent average before-tax ratio rose 0.035. The average after-tax profit-to-

TABLE 8.8

Percent by Which Before-Tax Rates of Return Exceed After-Tax Rates of Return

$$\left(\frac{[\text{Before-Tax Rates}] \text{ minus } [\text{After-Tax Rates}]}{\text{After-Tax Rates}} \right)$$

Year (1)	Total Mining		Metal Mining	
	Profit to Capital Employed (2a)	Profit to Equity (2b)	Profit to Capital Employed (3a)	Profit to Equity (3b)
1962	0.21	0.23	0.21	0.20
1963	0.23	0.21	0.20	0.20
1964	0.21	0.21	0.21	0.20
1965	0.19	0.21	0.23	0.23
1966	0.24	0.23	0.24	0.23
1967	0.20	0.22	0.23	0.23
1968	0.26	0.25	0.28	0.28
1969	0.29	0.27	0.26	0.26
1970	0.43	0.41	0.42	0.43
1971	0.22	0.23	0.23	0.22
1972	0.48	0.49	0.94	0.88
1973	0.38	0.40	0.52	0.51
1974	0.80	0.78	0.83	0.83
1975	0.78	0.82	0.87	0.86

capital ratio fell 0.007, while the average before-tax ratio increased 0.015. While the trend in metal mining was not so pronounced and consistent, it certainly appears that the effect of the change in taxation is being felt not primarily in a reduction in after-tax returns but in an increase in before-tax returns, at least in total mining.

Thus, it appears that the tax changes have (1) made the impact of taxation greater in mining, as evidenced by the increased difference between before- and after-tax returns, and (2) caused less profitable undertakings to be abandoned or cut down. The effect, of course, is a short-run phenomenon thus far, whereas our predictions about the effect on investment are long-run estimates. In any case, the results give some evidence that our analysis is certainly reasonable and accurate. Of course, the total impact on investment is an empirical question, but the substantial increase in the differences between returns lend credence to our range of estimates as to the severity of the impact.

FIGURE 8.4

Percent by Which Before-Tax Returns Exceed After-Tax Returns, Total Mining

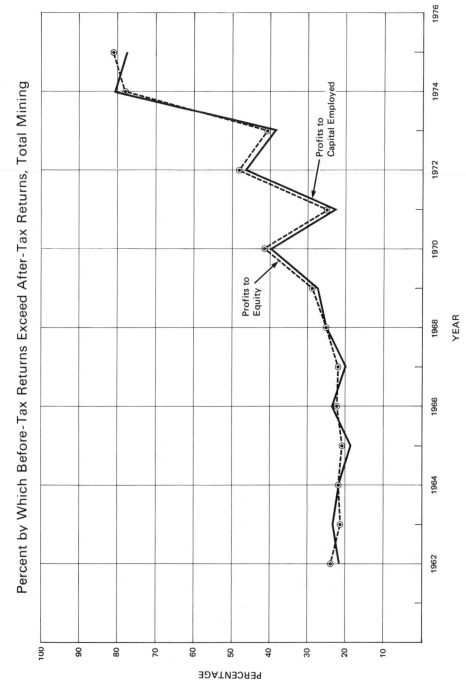

FIGURE 8.5

Percent by Which Before-Tax Returns Exceed After-Tax Returns, Metal Mining

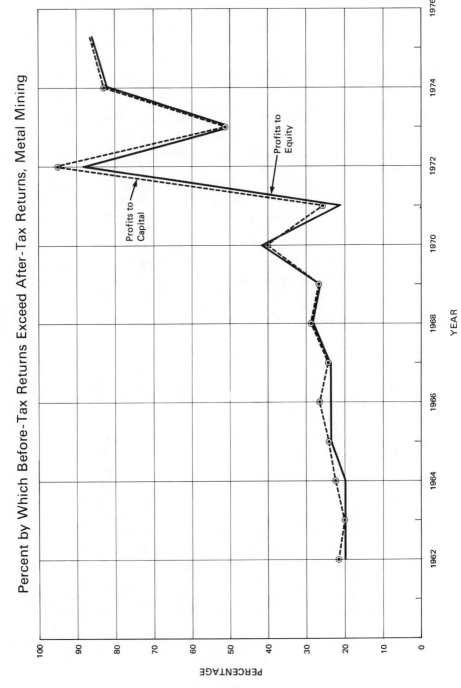

SUMMARY AND CONCLUSIONS

In this chapter we employed the approach developed in Chapter 7 to estimate the impact of changes in taxation or environmental controls on investment in the mineral-extracting industries. In very general terms, our model estimates the level of investment forthcoming relative to that which would have occurred in absence of the changes by comparing the present values of an investment before and after the changes.

In order to illustrate this estimation procedure, we used our model to estimate the impact of some actual policy changes that were introduced in Canada in the early 1970s. We employed estimates of tax rates, discount rates and the cost of environmental controls supplied by the Ministry of Natural Resources of the Province of Ontario and by private mining firms operating mines or smelters in the Province of Ontario. The value of a specific income flow under the tax rates, discount rates, and environmental controls that existed prior to recent changes was then compared to the value of the same income flow under the new set of parameters. Using these ratios, we then calculated a range of expected impacts caused by changes in effective tax rates and environmental restrictions. Our estimates indicated that the changes in tax rates and environmental controls should be expected to have a significant impact on the level of investment in the mining industry in Canada. Indeed our "best" estimates indicate that, as a result of these changes, total investment would be 25 to 43 percent less than that which would have occurred in the absence of the changes. Since sufficient time has not passed to ascertain the actual impact that has occurred, we sought to substantiate our projections by seeking to find an indication of the fact that (1) a disturbance had occurred and (2) the mining industry was adjusting, as indicated by growth in the deviation between before- and after-tax rates of return.

We showed that there is clear and irrefutable evidence that in the 1970s the deviation between before- and after-tax rates of return on investment in the mining industry in Canada has increased substantially, compared to similar figures for manufacturing. The before-tax rate of return in mining has grown out of proportion with a similar figure for manufacturing, while the after-tax rate of return has remained virtually the same. This is simultaneously indicative of two relevant factors: (1) The effective tax rate has risen, and (2) the mineral industry is adjusting to the new high rate by being more selective in its investment projects, foregoing investment opportunities at lower potential returns that would have been acceptable before recent changes in the tax rate.

In order to verify empirically the projections made concerning the impact of taxation and environmental controls on investment decisions, it would be necessary not only to have data over a sufficiently long time period in which such adjustments in investment could occur; we would also require information about what the investment would have been in the absence of such changes. This is information that the policy analyst, making a decision in the present that will affect the future, will never have. We believe, however, that the simple tools developed in Chapters 7 and 8 provide a substantial improvement over the current system of projection of impact that relies heavily on intuitive knowledge and background of the mining industry, which few policy analysts possess.

NOTES

1. This type of relation is exactly what was shown in a previous study using data up to 1971. See G. Anders, W.P. Gramm, and S.C. Maurice, *The Impact of Taxation and Environmental Controls on the Ontario Mining Industry – A Pilot Study* (Ontario Ministry of Natural Resources, Toronto, 1975).

2. See Chapter 3 for a complete description of these ratios. The ratios were computed from data in and definitions given by *Statistics Canada*. As mentioned in Chapter 3, we did not have access to profitability data prior to 1962.

9

An Analysis of
the Benefits from
Environmental Controls on
the Mineral-Extracting Industries

In this chapter we present an analysis much more rigorous than that in Chapter 6 of ways to evaluate the benefits forthcoming from the imposition of environmental controls upon mineral-extracting firms. Chapters 7 and 8 were concerned with the economic costs to society of such imposition. But unless people thought that these controls would lead to social benefits, there would be little discussion or argument concerning controls being good or bad. Neither would there be any argument over the magnitude or severity of environmental regulations. The only way that one can determine the usefulness and the optimal level of regulations is to balance the cost to society with the expected benefits.

The problem is that the measurement of benefits is a task of even greater magnitude than the estimation of costs. This is the case primarily because of the difficulty of obtaining data but additionally because of the unsatisfactory status of the theory for the estimation of pollution damages from mineral industries. This sketchy theoretical structure stems from the fact that mineral industry pollution does not follow the pattern of the "typical" theory of pollution damages set forth primarily to explain pollution from manufacturing in rather heavily populated areas. The types of externalities in each case are quite different.

In this chapter we set forth the theory of estimation of benefits from externally imposed environmental regulations upon mineral-extracting industries. The "typical" theory of externalities and benefits is first described and then compared and contrasted with conditions that generally prevail for mineral-extracting firms. It is shown that the typical theory does not fit mineral industries particularly well. We then modify the theoretical structure in order to analyze the situation that exists in the mineral-extracting sector. We next provide a methodology by which the economic effect of environmental controls can be considered empirically and illustrate this technique using data drawn from the Province of Ontario. We conclude this chapter with a discussion of the effects of environmental controls on the employees of the mineral-extracting industries.

THE "TYPICAL" THEORY OF EXTERNALITIES AND BENEFITS FROM CONTROLS

Frequently, in fact in the great majority of cases, externalities from pollution are discussed within the context of previously established property rights. However, the temporal assignment of these property rights is usually ignored. The traditional textbook example often takes the following form. A poor woman supports herself by doing hand laundry. Since she cannot afford an automatic dryer, she must hang the clothes in her yard to dry. A nearby factory emits smoke and soot that dirty the laundry, forcing the woman to wash again. The example generally goes on to point out that the problem is insoluable. Because the marginal cost of smoke disposal exceeds the marginal private cost, the factory creates an external diseconomy on the woman. The factory owner has no interest in buying the hand laundry because he then would be confronted with the smoke problem. The woman owns a scarce resource, the air space over her yard, but she cannot charge for its usage. Therefore, the factory uses it to her detriment.

The analytical problem is that this and similar cases are oversimplified in the sense of confusing the existence of pollution with the incidence of pollution damage. Consider again the case of the one-woman laundry and the smoke-belching factory. Certainly, the poor woman is being damaged by the pollution. But she may have purchased her home near the factory precisely because it happened to be in a polluted environment, and accordingly, she was able to buy it much more cheaply than she could similar homes in more desirable environments. Naturally, smoke controls would benefit the laundry at the expense of the factory, but this is not to say that the pollution is damaging in the sense of causing costs to be higher than they were in the planning stages.

On the other hand, if the laundry was there first, the smoking factory may not have been anticipated and to the extent that the value of the property falls, the woman suffers an uncompensated windfall loss. Alternatively, it is quite possible the person who suffered the loss was the owner of the property prior to the building of the factory, who then sold the property to the woman after the factory was built. In this case controls would benefit the woman, who suffered no net economic loss from pollution, but would not benefit the individual who actually suffered the loss—the person who owned the land prior to the installation of the factory.

The point here is that the benefit-loss criterion concerning who harms whom is not quite so straightforward as is frequently assumed. In other words, pollution may well cause damage to those affected by it, but, because of the workings of the market, those damaged are compensated in the form of lower costs or in some other manner, such as increased compensation. To be sure someone may well have suffered damage in the form of a reduction in wealth, but the one damaged may quite easily differ from the one who would benefit from environmental controls that improved the environment.

This divergence between those damaged by pollution and those benefitted by the installation of controls is of utmost importance in analyzing the benefits of controls in mining industries. Environmental control legislation in mineral industries frequently applies equally to practically all areas. Unfortunately, the law often does not differentiate between firms whose pollution does not result in economic damage and those who cause damage. Some firms are legally required to reduce emission even though there is little *net* benefit to any resource owner. The abatement of pollution does not guarantee the existence of net benefit. Furthermore, where economic damage is internalized through negotiation, the control can result in a nonoptimal amount of damage. Let us focus now on damages and benefits in mineral extraction.

EXTERNALITIES AND BENEFITS FROM CONTROLS IN MINING

There are three basic differences between the typical theory of benefits from controls and that in the case of mineral extracting industries. Frequently in the latter case, but much less often in the former, (1) many of those affected by pollution are associated with the polluting firm in additional ways; (2) many of those damaged established their property rights after the damages or in full knowledge of the existence of potential damage; (3) the damage is generally quite localized.

To simplify our analysis we divide those affected by pollution, and hence by controls, into two groups. The first is made up of the owners of physical capital, particularly land. The second group is made up of those whose human capital is affected. We consider an area in which one or more firms operate mines and smelting operations that, in the absence of controls, have been polluting the environment to some extent.*

----

*We certainly do not imply that, in the absence of controls, firms would take no action to abate pollution. There are, of course, several motives for firms to abate voluntarily. As will be

Owners of Physical Capital

First, we turn to the effect upon the owners of the land leased by the mining firms. If the leased land is privately owned, we would assume that the owner recovers any losses expected from pollution, such as the value of lost trees or impure streams in the leasing arrangement. Obviously, no private owner would have leased the land if the use did not go to the highest bidder. Thus, the land would go into its highest valued use, and the returns would be sufficient to compensate the owner for the lost value of the resources due to pollution.

If the leased lands are owned by the government, the lands still go the highest-valued use, providing that government attempts to maximize revenue. If, however, government "gives away" leases or if it does not recover the expected loss of resources, then the value of the pollution loss will not be recovered in the leasing arrangement. Even when the government receives its returns through taxation, the effect is in large measure similar to the private arrangement. In any case, if government acts economically, the leasing arrangement should cover some of the cost of the lost resources.

If, when it made the leasing arrangement, the mineral-extracting firm did not expect additional controls, the future cost involved in such controls would not have been taken into account originally, and the imposition of controls would involve a windfall gain to the landowners. But, generally the loss and the gain would not be exactly offsetting. Furthermore, the windfall gains to the landholders would not always be without some offsetting costs to them.

The situation is relatively unchanged when the land is owned by the mining firm (or firms). In this case the firm suffers the costs, but it also benefits from the controls as landowner. It must generally be the case that the firm is a net loser under imposed controls, because the firm would have been motivated to institute the controls with no external pressure if it could have benefitted from such imposition.

There is, however, the so-called free-rider problem for cases in which several firms in an area do the polluting. That is, although no one firm could benefit if by itself it reduced its own pollution, because the damage from all other firms' pollution would continue, if all firms decreased pollution together, each would benefit to a greater extent than the increase in costs. In such a situation it is highly unlikely that pollution will be reduced if firms act independently.[1]

---

described in the section on benefits, the firm may wish to pay its workers income in kind through better working conditions. Also, there could be pure civic conscientiousness. We simply wish to analyze the effect of controls that call for increased abatement.

Thus, we might say that the owner of the leased lands would benefit from controls to the extent that the value of the land increases. But, if part or all of the expected damage was accounted for in the leasing arrangement, the owner is doubly compensated and the mine is doubly charged. If the damage was not accounted for in the leasing arrangement, no matter whose fault, the owner benefits to the extent of the increased value of the land less the reduction in returns because of reduced royalties or taxation. In any case, all of the circumstances should be taken into account when estimating the economic benefits of pollution in the case of landowners.

We next turn to the owners of adjacent lands and properties. Those who own property affected by the pollution but are not compensated through the leasing arrangement are in a situation similar to that of the woman with the laundry near a factory. The key point is whether or not ownership began prior to the polluting. If the property was purchased while both buyer and seller knew the probability of pollution damage, then the damage costs were accounted for in the purchase price. Again, only the owner prior to the polluting process suffers a windfall loss. In the case where the original owner is not the current owner, environmental controls will doubly compensate for damages, because the purchase price would have taken into consideration the damages.

Of course, if both the leased property and adjacent property were government lands, the leasing arrangement would have taken the damages into consideration. If this were not the case, then the controls would involve net benefits to the landowners, but not in excess of the total damages.

As has been shown, the net effect of environmental controls upon property owners depends upon the conditions of ownership. The increase in value—generally used to measure the benefits from abatement—may well give an inaccurate picture of the total effect, particularly in the case of mineral lands and adjacent lands. The gross reduction in property values from pollution in the absence of controls in many cases would not measure the true loss to the property owners. If these losses were taken into account in the leasing arrangement or if property owners acquired the lands after pollution began, the owners would have been compensated for the loss prior to the imposition of controls. Thus, the increase in property value after controls would involve double compensation. Only in the case in which the owners prior to pollution and after imposition of controls were the same and there was no compensation involved, would the increased property value merely offset the loss caused by pollution.

Figure 9.1 summarizes the effects of controls upon owners of physical property not compensated by the mineral-leasing arrangement. The three situations are determined by the conditions of ownership before and after the beginning of the polluting activity. In the first situation in which the property

FIGURE 9.1

Effects of Environmental Controls on Owners of Physical Capital in Various Situations

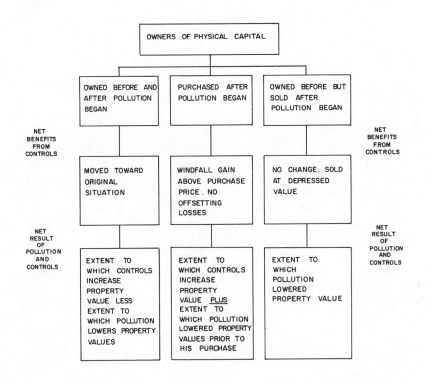

owners were owners before and after the beginning of pollution, the controls
to a greater or lesser extent offset the losses from pollution.

In the second situation the property was acquired after the beginning of
polluting activity. Thus the acquisition was at a lower price, reflecting the
lowered property value. Any benefits from controls would be a net windfall
gain. Someone who owned the property prior to the polluting activity but sold
afterwards at the depressed value would suffer a net loss from pollution but
would not be compensated by the controls. Thus, the net effect depends upon
the conditions of the property rights.

Owners of Human Capital

Certainly, air and water pollution and pollution abatement affect persons
besides those who own land or other physical property. People living in the
area can suffer physical damage from pollution. These types of losses and
benefits are conceptually much more difficult to estimate than in the case of
physical property.

The problem here is in the estimation of damages. Certainly, as we argued
in Chapter 6, one could obtain some estimate of the cost by determining how
much individuals would have been willing to pay in order to reduce the
pollution by certain amounts. This task would, however, be extremely difficult
to carry out in reality. Certainly one would give a different answer to the
question about the personal loss from pollution depending upon whether or
not compensation would actually be paid. Thus, the actual task of measurement would be overwhelming.

Aside from the measurement problem there is also the problem of *when*
the affected individual came to the polluted area. Just as in the case of owners
of physical property, the establishment of property rights determines in large
measure the net benefit from controls. First, consider those who lived in the
area prior to the beginning of the polluting activity and who remained. The
controls would have the effect of moving the individual to the original
situation. The basic loss is the effect occurring after pollution but before
controls. The total effect depends upon the extent of controls. Individuals who
move to the area after pollution began must have been compensated to some
extent to induce a move into a less desirable situation. That is, firms hiring
workers in the area would have to pay higher wages, the less desirable the
environment. Thus these residents must have already been compensated to
some extent for the disutility of living in the area. Otherwise they would not
have moved into the area. Unless controls are followed by a decrease in
compensation, these individuals are benefitted when there is no uncompensated loss. (We shall return to analysis of this type of situation below when
dealing with the effect of controls upon employees of the polluting firm.)

FIGURE 9.2

Effects of Environmental Controls on Owners of Human Capital in Various Situations

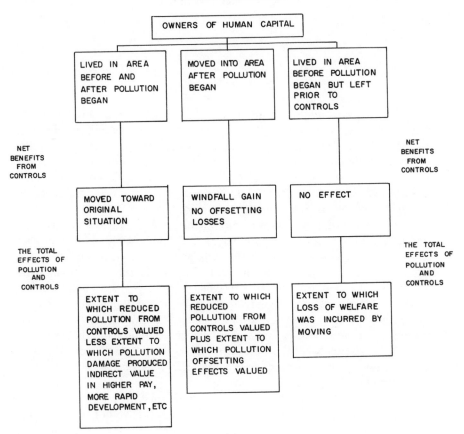

Finally, there are those residents who lived in the area prior to the pollution but then moved. The controls cannot under these circumstances compensate for the disutility of moving. These individuals, therefore, suffer a net loss from pollution, the total loss being the value attached to living in a less desirable situation. These effects are summarized in Figure 9.2. The analysis is similar to that used for Figure 9.1.

AN APPROACH TO MEASUREMENT OF THE EFFECT OF CONTROLS

We turn now to the estimation of the benefits from environmental controls. In the first sections of this chapter we set forth a simplified version of an elaborate and, we feel, proper method of measuring the economic benefits of environmental controls upon mineral extracting firms and industries. We argued that the mere counting of resources saved, etc., would generally overestimate the actual benefits. The actual benefits must involve differences in returns for owners of resources, such as land or labor.

To estimate the basic impact of controls, one would like to carry out simulations like those developed in Chapters 7 and 8. We could assume that controls such as those described in those chapters and are now in force were in fact enacted earlier, for example in 1961. Then we could, using engineering data and our theoretical structure, predict the impact upon environmental variables. If we knew all relevant supply-and-demand conditions, we would then estimate the benefits along with the effect upon wage rates, land values, human capital, and so forth. These overall *net* benefits could then be compared with the *net* social cost of the controls in order to estimate the net effect over time had these controls actually been imposed in 1961, or in any other year in which we chose to begin simulation.

Unfortunately, the data for such simulations are simply not available. The measures undertaken in response to recent controls have not been in effect sufficiently long to generate the engineering data. Furthermore, the total impact of pollution reduction on human and property values cannot be estimated or even approximated with the available data. Therefore, if we are to show the extent to which past controls, had they been enacted, could have benefitted society and to predict the future benefits from present controls, certain conceptual experiments are our only means of approach. We can make certain assumptions about the effects of controls and then, using the available data, deduce the total benefits. If the assumptions are somewhat exaggerated, then the benefits will be exaggerated also. But even predicting exaggerated benefits under exaggerated assumptions will at least give some indication of the bounds of the benefits that would be deduced under more "reasonable"

assumptions. In the absence of additional data this is about the best one can
do.

In this section we will illustrate this procedure by performing some of
these conceptual experiments. We will use in these experiments data drawn
from the Province of Ontario. Similar procedures could be employed for any
other region or country.

The assumptions under which we make our deductions are admittedly
quite simplified. Others may wish to make entirely different assumptions to
predict the range of economic benefits. Or they may wish to focus upon other
variables. In any case, the assumptions and methodology used here at least
provide a first approach to benefit estimation.

Estimation Assuming All Affected Area Is Transformed into Agricultural Land

For our first conceptual experiment, assume that the benefit from
environmental controls is in the form of land that would have been "disturbed"
by mining being turned into agricultural land. That is, suppose that in the past
a certain amount of disturbed land had not been disturbed, because of
controls, and had been used for agricultural purposes. In 1974 the Ministry of
Natural Resources of the Province of Ontario estimated that at that time
66,873 acres of land in the entire province were being "disturbed" by the
mining industry. This disturbed acreage included buildings, roads, waste
disposal areas, and any other land that would be different in the absence of
mining. According to sources in the ministry, this estimate, if it erred at all,
erred in the direction of an overestimate rather than an underestimate of the
actual disturbed area.

Let us now make the following assumptions. First, let us assume that the
total disturbed area in 1961 and in subsequent years would have remained
constant at 66,873 acres in the absence of regulation and that, because it was
disturbed, it had no economic use other than in mining. Next, assume that
environmental controls along the line of those in effect now were imposed
throughout the province in 1961. The hypothetical controls have the effect of
causing some land that would have been otherwise useless (in the absence of
mining) to become undisturbed. Finally, assume that all land that would have
been undisturbed, under our assumption, because of controls became
agricultural land of average productivity. In effect we assume that the
hypothetical controls turned disturbed land into agricultural land.

We wish to estimate the added *gross* receipts that would have been
forthcoming under these assumptions under several measures of the number
of acres that would have been changed into agriculture because of controls.

According to *Ontario Statistics 1975*,[2] the total provincial acreage devoted to agriculture during selected years was as follows:

Year	Acreage
1961	18,578,507
1966	17,826,045
1971	15,963,056

From the same source, total gross receipts in the province from agriculture in 1971 were $1.43 billion, a 63.6 percent increase from 1961. Extrapolating for 1966, the gross receipts in selected years were:

Year	Receipts ($)
1961	874,000,000
1966	1,152,000,000
1971	1,430,000,000

Using these figures the average receipts per acre used in agriculture in Ontario were, in the specified years:

Year	Average Receipts per Acre ($)
1961	47.04
1966	64.62
1971	89.58

Now let us make several assumptions about the amount of "disturbed" land that would have been made "undisturbed" and devoted to agriculture had controls been in effect over the period 1961–71. We use the following estimates of acreage converted to agriculture: *all* the disturbed land (66,873 acres); half of the disturbed land (33,436 acres); one-quarter of the disturbed land (13,718 acres). Finally, we posit that the 66,873 estimate is too low, and we double the disturbed estimate, so that 133,746 acres would be added to agricultural usage. The highest estimates are rather farfetched, but they do allow an estimate of the upper bound of possible benefits from controls as measured in agricultural land use alone.

The hypothetical additions to gross agricultural receipts under the various assumptions along with the percentage increase in gross receipts for the various years are shown in Table 9.1. In no year did gross agricultural receipts increase as much as 1 percent, even for the very highest estimate of affected land area. In our experiment, if every acre of the disturbed land had gone into average agriculture immediately, only $3,146,000 would have been

TABLE 9.1

Simulated Additional Agricultural Receipts under Environmental Controls

	1961		1966		1971	
Added Agricultural Acreage (1)	Percent Increase in Receipts (2)	Additional Agricultural Receipts ($1,000) (3)	Percent Increase in Receipts (4)	Additional Agricultural Receipts ($1,000) (5)	Percent Increase in Receipts (6)	Additional Agricultural Receipts ($1,000) (7)
13,718	0.09	587	0.10	1,095	0.11	1,502
33,436	0.18	1,173	0.19	2,189	0.21	3,003
66,873	0.36	3,146	0.38	4,378	0.42	6,006
133,746	0.72	6,293	0.76	8,755	0.82	12,011

added to the total *gross* receipts of $874,000,000 in 1961, an addition of 0.36 percent. Even by 1971 the total receipts of $1,430,000,000 would have been increased by only $6,006,000, an addition of 0.42 percent.

We should emphasize that these estimates are *gross receipts* from agriculture; they are not net of other costs. Neither do they indicate that the total product of the Province would increase by that amount. Other resources would have been given up to develop the agricultural productivity of the land. In any case, land use in agriculture would be the absolute maximum value of the released land. Therefore, these figures probably are significantly greater than the actual economic benefits that would have been forthcoming. Clearly these maximum benefits as measured in agricultural land addition are quite small relative to the total costs. Furthermore, the amount of disturbed land that would have been undisturbed under controls has also probably been significantly overestimated.

Estimates Assuming All Affected Area Goes into Forestry

Lumbering rather than agriculture is probably a more realistic alternative use of disturbed land. Therefore, let us use our previous estimates of land that would not have been disturbed if controls had been in effect and assume that land would have been used for lumbering during the period.

In 1973 there were 92,297,000 acres of primary forest land in the Province of Ontario. Using the same three years as before, 1961, 1966, and 1971, the volume of production, the value of shipments by the forestry industry (own manufacture), and value of shipments by the logging industry are as follows:*

Year	Volume of Production (mm cu. ft.)	Value of Shipment, Own Manufacture ($1,000)	Value of Shipment Logging ($1,000)
1961	494	$1,094,837	$147,060(1963)
1966	601	1,566,252	190,784
1971	559	1,934,406	206,608

Next let us assume that, because of environmental controls, certain portions of the disturbed land were added to primary forests and that these portions were forested to the same degree as total primary forests. For these estimates we use the above estimates of the land freed by environmental controls and transferred into agriculture: 13,718; 33,436; 66,873; and 133,746 acres.

Assume for simplicity that in each year the best estimate of primary forest land in Ontario was 92,297,000 acres, the estimate for 1973. Table 9.2 column 2 shows the percentage increase in primary forest acreage for each of the four estimates of land made "undisturbed" by controls, shown in column 1. Columns 3, 5, and 7 show the *additional* value of total forestry shipments, assuming that the "freed" acreage yields the average returns from all primary forest land for the years 1961, 1966, and 1971, respectively. These added returns range from $218,967 in 1961 for the 13,718 estimate to $2,708,168 in 1971 for the 133,746 acre estimate of undisturbed land. All of the estimates are small, particularly when compared with the capital expenditure needed to turn these acres from disturbed to undisturbed land. Columns 4, 6, and 8 show the hypothetical additional logging receipts for each year under the various assumptions. Obviously, these are much smaller than the additional lumbering receipts.

It would appear that the only other alternative use of these freed lands would be to turn otherwise disturbed lands into hunting areas.† Let us say that 0.07 percent is the average increase in primary forest acreage from controls. If this added acreage increases the average moose and bear kill in Ontario by about the same percentage an additional 9.1 moose and 6.3 bears would have been killed per year. Given that moose hunters spend only about $11 million

---

*The value of shipments for logging in Ontario is not available for years prior to 1963. Hence, we employed the 1963 data as an estimate for 1961.

†Since approximately 75 percent of the Ontario fishing industry is involved with the Great Lakes, we did not examine the impact of controls in fishing.

TABLE 9.2

Simulated Additional Receipts in Either Lumbering or Logging under Environmental Controls

Added Forestry Acreage (1)	Percent Increase in Forests (2)	1961		1966		1971	
		Additional Total Receipts ($) (3)	Additional Logging Receipts ($) (4)	Additional Total Receipts ($) (5)	Additional Logging Receipts ($) (6)	Additional Total Receipts ($) (7)	Additional Logging Receipts ($) (8)
13,718	0.0002	218,967	29,412	313,250	38,157	386,881	41,321
33,436	0.0004	437,934	58,846	626,500	76,313	773,762	82,643
66,873	0.0007	766,386	102,942	1,097,376	133,523	1,354,084	144,625
133,746	0.0014	1,532,771	205,884	2,192,752	267,047	2,708,168	289,251

and bear hunters $2 million annually, this alternative use of disturbed lands would not seem to add significantly to the total monetary product of the Province of Ontario.

Summary and Conclusion

Even using the largest estimate of undisturbed land that would have been disturbed and the highest conceivable value for alternative land usage, the increase in the total provincial production from land use produced by environmental controls would have been extremely small. The more likely estimates of land and alternative uses lead to estimates that are minute, particularly when compared with the expenditures on environmental controls necessary to achieve the results. Moreover, we ignored the fact that an increase in agricultural or forestry returns would have drawn resources from other uses. Such withdrawal would have decreased output in other areas, causing the impact to be even lower.

On the other hand, we may well have ignored an important benefit of environmental controls in remote mining areas. These are those individuals who receive considerable psychic income or personal utility from knowing that there is lowered pollution in Northern Ontario. It is possible that the distaste these people may have for pollution is as real a cost to them as any lost economic value for those whose physical property is damaged. These are the benefits that are difficult, perhaps impossible, to estimate. Yet if our estimates of the potential economic benefits of environmental controls are anywhere close to accurate, the imposition of controls must be justified on psychic grounds. The pure economic gain from controls, even using the largest reasonable estimates, is potentially so low that the cost of controls must far exceed the economic value of the benefits. Thus, if a case is to be made that the benefits of controls outweigh the costs, the case must rest on these psychological or nonpecuniary returns. While these noneconomic benefits are quite real, in the sense that they affect the utility of individuals, it is beyond the scope of this analysis to measure these benefits monetarily, even though it would be conceptually possible to do so.

Therefore, all we have done here is to make some simplifying assumptions that permit us to estimate a range of potential economic benefits of environmental controls on Ontario mineral industries. These benefits can be compared with the cost. Others may wish to make entirely different assumptions about the amount of disturbed land, the extent to which controls would leave some of these lands undisturbed, and the alternative use of the affected lands. While there is clearly a need for further empirical evidence, the assumptions and methodology used here at least represent a first approximation to the economic benefits.

BENEFITS TO EMPLOYEES OF CONTROLLED FIRMS

We have thus far neglected the effect of environmental controls on a proportionally large group of individuals, the employees of the mineral firms. Since a large percentage of some of the communities consists of employees and their families, this is a potentially important group when estimating the benefits of controls. But because, for reasons to be shown below, employees are affected quite differently from other parties, the benefits to this group are considered separately.

Many types of environmental regulations affect employees of mineral firms. In addition to controls on air and water emission, some of these are noise controls, dust controls, and other in-plant regulations. We begin with a brief description of the theory of the job choice, then we apply this theory to an analysis of the effect of regulation upon employment, wages, and the environment of the workers, showing why the effect upon employees differs from the effect upon other affected groups.

The Environment and Wage Rates

To begin the analysis, consider a mineral industry composed of many somewhat similar firms. None of the firms is at first subject to any type of environmental regulations. Of course, any effective environmental regulation must cause a firm—or all firms—to take some course of action that would not have been taken in the absence of the regulation. But the absence of formal regulations by no means implies that the firms themselves take no actions to alleviate irritative dust, noise conditions, or other environmental disturbances. Certainly, we must note that firms historically have in fact taken a considerable amount of action in the interest of improving working conditions. These remedial activities were not necessarily out of the goodness of management's hearts—although some action may have been partly or even solely philanthropic in nature. Let us, therefore, first examine some motivations for these activities that are a result of strictly market-oriented forces. To this end we need to use the economic theory of job choice.

The best way to introduce the economic theory of job choice is to begin with a rather old but famous joke. It is told that George Bernard Shaw was seated next to an attractive, sophisticated lady at a dinner party. Making small talk with the lady, Shaw asked her whether she would consider sleeping with him that night for the payment of one million pounds. The lady thought about it then said that yes she probably would. Shaw then asked her if she would spend the night with him for one pound. The lady answered, indignantly, "Sir, what do you think I am?" Shaw replied, "Madam, we have already settled upon what you are; now we are merely haggling over price."

Shaw may have been quite clever in his retort, but he was exhibiting a lack of economic understanding. As an analogy, we might ask someone whether he would haul away some garbage for $100. Probably most persons would; however, most would probably not do it for ten cents. This situation would, by no means, classify those persons as garbage collectors. Neither should the lady be classified as what Shaw classified her solely on the basis of the conversation.

The point of the story and the basis of the theory can be expressed systematically with a simplified model. Assume an economy in which there are only two occupations, designated as A and B. Members of this economy have varying degrees of preference between the two occupations. For some it would take a very high wage in B relative to the wage in A to induce them to work in B rather than A. The opposite applies for those who strongly prefer B to A. On the other hand, for those who are almost indifferent about choosing between the two occupations a slightly higher rate in A induces employment in A, and vice versa. In any case an equilibrium wage ratio is attained, and the economy's employment distribution is determined.

The situation is shown graphically in Figure 9.3. The ratio of workers in A to workers in B is plotted along the horizontal axis; the ratio of the wage in A relative to the wage in B, along the vertical. Again, there are only two occupations. The downsloping demand shows that as the wage in A falls relative to the wage in B (W_a/W_b declines), society demands more workers in A relative to workers in B (N_a/N_b increases).

The upward sloping supply curve shows that, as the wage in A rises relative to the wage in B (W_a/W_b rises), the proportion who wish to supply labor services to A rather than B rises also (N_a/N_b increases). As W_a/W_b decreases, N_a/N_b decreases also. Of course the position of the supply curve depends upon all of the conditions of employment in each occupation. In Figure 9.3 with the supply and demand situation shown, the equilibrium wage ratio is $\overline{W}_a/\overline{W}_b$; the ratio of employment then is $\overline{N}_a/\overline{N}_b$.

Suppose now that something occurs to make work in occupation A more desirable than before. Possibly the working conditions in firms hiring A workers are substantially improved. When workers find out that these new conditions are more desirable than previously, some workers will at the original wage ratio transfer out of B and try to find employment in A. This means an increase in the supply of A workers and a decrease in the supply of B workers. This shift has the effect of driving down the wage in occupation A relative to the wage in B. A new equilibrium is determined.

In terms of Figure 9.3, the newly increased desirability of employment in A causes the original supply to increase to the new supply designated by the dotted line. This shift in supply means that at each of the wage ratios, W_a/W_b, a larger proportion of the work force wishes to supply its services to A rather than B (N_a/N_b increases). Clearly, this shift in supply causes the wage in A to fall relative to that in B and employment in A to rise relative to that in B. Firms

FIGURE 9.3

Ratio of Workers in Occupations A and B

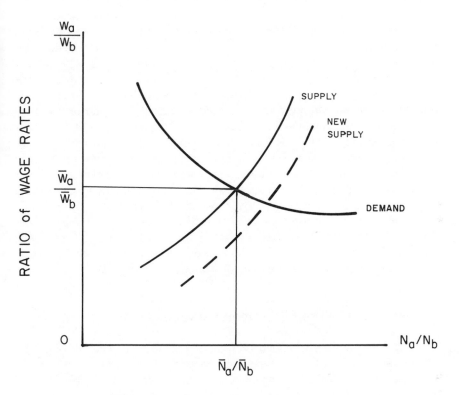

hiring A workers now can get more workers less expensively, but one cannot say whether these firms are better off. After all, the improvement in working conditions may not have been completely costless. Some comparison of the costs and benefits must be made.

Alternatively, if working conditions in occupation A deteriorated, employers of A labor would have to pay higher wages than before in order to induce labor to work in that occupation. We have concentrated upon shifts in supply induced by changes in working conditions. We could just as easily have examined changes in demand occasioned by society's wanting more or less of a specific occupation at each wage ratio in the range. For our purpose—an examination of the effect of environmental controls on labor—we need only the supply shift.

There is no real contradiction between our theoretical structure and what we observe, even though on first impression some may note an apparent contradiction. Our theory implies that higher wages must be paid to induce people to enter less desirable occupations, and the more desirable jobs can be filled at lower wages. Of course, what is desirable to one may not be desirable to another. We observe, however, that less desirable jobs, such as garbage collection, generally pay less than do more desirable positions, for example, being president of a corporation or a medical doctor. But note that there are differences in the supply offered to various occupations. There are great disparities in the qualities employers wish in these jobs, in the academic training required, and in the legal ability to enter certain occupations. While purely manual labor may, to most people, be undesirable, if a large number of people cannot, in the eyes of society, do much else (or are not demanded at the going wage rate in other occupations), the large supply of manual laborers, or those working at such jobs, will drive down wages in that sector. Being a brain surgeon, a corporate executive, or a movie star may be desirable, but the training and attributes required along with the limited number of applicants who are accepted keep wages high in these occupations. Therefore, in all cases both supply and demand must be considered.

The Effect upon Individual Firms

We can use the above analysis to determine the economic impact of in-plant regulation and other types of environmental regulations that affect workers. We can see easily why a plant such as a mine or a smelter would voluntarily install such features as noise and dust controls. To induce additional workers away from other jobs or from nonemployment, a firm has several choices. It can bid for workers through the incentive of higher wages, it can improve working conditions by environmental controls, among other things, or it obviously can, and probably in most cases does, use some combination.

No profit-maximizing firm would knowledgeably spend money upon improving working conditions if it could attract the same types and number of workers more economically with higher wages, and vice versa. In other words, firms would balance the marginal returns and the marginal expenditures from investment in in-plant environmental controls (and other changes in working conditions) and from changes in the wage rates. Of course, the effect of environmental improvements—particularly in-plant improvements—upon worker productivity and tenure of employment would be weighted in the analysis of returns also. Firms would invest in such improvements even though they believed that such investment would have little or no depressing effect upon wage rates as long as the marginal value of the productivity and tenure improvements exceed the marginal cost of the improvements. For all firms in the industry there would result some sort of equilibrium-employment package.

Under some circumstances environmental improvements might be undertaken even though management does not believe that the added cost of such improvements will be offset by the decreased wages or the increased labor productivity. If the firm is making above-normal profits or rates of return in excess of the going rate on capital, it may indeed wish to take some of these returns away from stockholders and give them to workers in the form of goods in kind: improved working conditions. This type of allocation cannot be done under reasonably competitive conditions, however. Competition forces returns to the going rate. When a reallocation of expenditure drives profit below the going rate for extensive periods of time, the firm finds its sources of capital disappearing, and in the long run when normal profits—meaning the going rate of return—are not made, the firm goes out of business. Both in the case of doing "good" (for example, improving working conditions more than is economical) and in doing "bad" (for example, discrimination against productive workers of a particular sex or race when the discrimination costs profits), the deviation from "normal" actions can force firms out of business, if these deviations drive profit below normal rates of return. A Firm earning returns well above normal, abstracting from risk, can do both, but only at a cost to itself or the stockholders.

Governmentally regulated environmental restrictions have an effect similar to the voluntary improvements that are uneconomical. These regulated improvements in environment benefit the firm to the extent that they enable firms to pay lower wages than would have been the case and to the extent that worker productivity is improved. The regulations lower the net returns to the firm to the extent that they increase costs above the level that would have prevailed otherwise. If the firm did not carry out the improvements voluntarily, then the net effect upon the firm would have to be a reduction in profits.

The point of our analysis is, however, that one should certainly not infer that the firm or the stockholders are the only ones who pay for such environmental improvements. Workers pay also in the form of wages lower than would have been obtained otherwise. For example, if a firm hires 100 workers at $25 a day and spends $5 a day for each worker in order to adhere to the environmental regulations, the firm's total wage bill is $30 per day per worker. If the $5-per-day environmental restrictions were not imposed, working conditions would be worse and the market would require the firm to pay higher wages in order to induce workers to work under the less amenable conditions. The firm could at the limit be forced to pay as much as $30 a day— the total amount it pays under regulation—in the absence of regulation. To the extent that workers value environmental improvement more than the reduction in wages, they are better off under regulation. To the extent that they value the higher wages more, they are worse off. In any case, to a greater or lesser degree, workers do pay something for improvement under the regulations.

The inclusion of labor unions does not alter the analysis extensively. No union can get the wage rate so high that the firm earns below the normal return over any extended period of time without creating a loss of jobs. Certainly, both unions and firms recognize that improved working conditions, including environmental changes, are in some measure a substitute for increased wages. During negotiations, both areas are discussed, and the settlement is announced in the form of an entire contracted package rather than as pure wage rates.

To summarize, when an environmental regulation that improves workers' environment is placed upon an individual firm—or mine or smelter—it will raise the costs of the firm, to the extent that the firm would not have done on its own what is required by regulation. If the firm is earning above-normal profit, the added costs will in some measure be borne out of profits. If a firm cannot in the long run earn the going rate of return, it will go out of business. Finally, to the extent that working conditions and wages are substitutes in the entire wage package to workers, the workers will absorb some of the cost of improved conditions in the form of wages below what would have been the case in the absence of regulation.

There is one thing the firm cannot do unless it has a significant amount of power in the market in which it sells the commodity. The individual firm, acting alone in response to a regulation imposed solely upon itself, cannot pass along the cost of the regulation in the form of increased prices to purchasers of its product. The competitive firm is a price taker in the product market. Capital and labor must bear the cost in this case. However, a regulation imposed upon an entire industry may involve different circumstances.

The Effect upon Industries

Let us assume now that the environmental regulations are imposed upon all of the firms mining and/or smelting a particular mineral in a particular region or even throughout the entire country. We could easily go so far as to consider a regulation affecting mining and or smelting firms for all minerals. When analyzing the effect upon workers, these assumptions lead to basic differences from our analysis of a regulation solely upon an individual firm.

First, since any one firm would not be likely to possess market power significant enough to affect price, it could not pass along the costs of the regulation to consumers in the form of increased prices. That is, in most instances one would think that any individual firm could increase or decrease its own output without having any effect upon the world price of the mineral. But it may be the case sometimes that if all firms extracting a particular mineral in a region or in a country increase or decrease their rate of output together, the world price would be affected.

The environmental regulations have two immediate effects. They improve working conditions and increase the production costs of all firms. But the improved working conditions increase in a greater or lesser degree the available supply of labor to the portion of the industry affected by the regulations. This can have some dampening effect upon wages, which may well offset to some extent the increase in production costs. Thus, labor pays some portion of the improvement. But the increase in labor supply will not be so marked as in the case of a single firm. The individual firm that improves environmental conditions can draw workers from other firms in the industry as well as from outside the industry. If a large segment of the mineral industry improves conditions, the possible increase in supply is much more limited. Thus, wages will be somewhat less affected.

Furthermore, if the output of all firms in the industry has some impact upon the world price of the mineral, the increase in production costs, to the extent that it is not offset by wages lower than would have existed in the absence of regulation, decreases industry supply, which, depending upon the elasticity of demand for the product and the importance of the regulated sector of the industry, causes the world price of the commodity to rise. In this way some of the cost of regulation is passed on (shifted) to consumers.

If some marginal firms find that the commodity price does not increase sufficiently to offset the increased costs and consequently cannot make a normal return, they will exit in the long run. Furthermore, those prospective investments that would have been profitable in the absence of regulation but are no longer economical will be foregone. In any case, the return to capital in the affected industry will return to the going rate for the economy. Changed rates of investment responding to the changed profitability conditions will cause this equalizing effect, as we have noted frequently. Thus, to summarize, the cost of the regulation will be absorbed in some degree by customers in the form of increased prices, and by labor in the form of decreased wages. Those firm owners who go out of business and the capital owners who suffer decreased returns will bear some of the costs in the short run. In addition, those firms earning above-normal returns in the form of rents will absorb some of the cost increase. However, as we have emphasized so frequently, in the long run the return to capital will be driven to the going rate of return for the economy.

The analysis is changed somewhat if the affected portion of the industry is a price taker in the world commodity market. Assume that the industry is approximately in long-run equilibrium with some firms making somewhat more than the alternative cost of capital—generally because of better ore bodies—and some marginal firms are making only the going rate of return. Again, let an environmental restriction that improves working conditions but adds to costs be imposed upon all firms.

In the same way as in the above analysis, the first effect of the regulation is to increase the supply of labor to the entire industry, because the improved working conditions make employment in such occupations more desirable. As indicated above, the increased supply will cause long-run wages to be lower than they would have been in the absence of regulation, but as indicated previously, the impact upon each firm in the form of increased labor is less when every firm in the industry is regulated than when only a single firm or a few firms are affected. In any case the regulation does cause some of the cost to be covered by reduced wages.

While some of the increased cost will be offset by the increased supply of labor, the expenditures will increase the costs of production beyond the level that would obtain the absence of such regulation. Certainly, if the increased costs are substantial, firms on the margin of profitability would in the long run exit the industry. This shutting down would, to some extent, free workers and increase the supply of labor to other firms, thereby causing additional downward pressure upon wages. In addition, firms contemplating entering the industry or even established firms considering new investments could find that the added requirements turn an investment from potentially profitable to insufficiently profitable—*insufficient* in the sense of not yielding the opportunity cost of capital. Thus far the analysis is similar to the case described above when the industry can affect the world price.

The important difference here is that in the case of an industry that is a price taker in the world market none of the higher costs can be passed along to purchasers in the form of higher prices. The increase will be borne internally. Some part of the cost of improved working conditions can be shifted in the form of wages lower than would be the case without regulation, although, certainly, the presence of labor unions may lengthen the readjustment process. Of course, any firms that are driven out of business will lose also. Finally, some of the costs will be absorbed by intramarginal firms, but investment even by these firms would be lowered by the increased costs.

Who benefits? Obviously labor would benefit if the increased utility from these improved working conditions more than offsets the cost to them—out of wages—of the changed conditions. The extent of the effect is an empirical question. Certainly, the cost to labor will be less when some can be passed on to the world market in the form of increased prices.

Fringe Benefits versus Wages

One theoretical conclusion in the case of environmental regulations is that those who benefit from such regulations in some part are those who pay for them. The net benefit-cost balance depends upon the responsiveness of labor supply and the subjective evaluation of those who benefit from the changed conditions. The question is, how valid is our theory? Do owners of resources in fact bear much or even some of the costs of regulation? It would be difficult, but probably not impossible, to measure accurately the portion of costs actually borne by resource owners.

We do, however, have available another method of approach: We can see whether our theory—which says that labor, among other resource owners, bears some of the cost of compliance—is consistent with what we observe in reality. To this end, consider a union negotiating a labor contract. We would expect the union to bargain over two items: wage rates and fringe benefits, including working conditions. One might ask why a union would wish additional fringe benefits if in fact they cause wage rates to be less than they would have been in the absence of such benefits. Of course, if nonmonetary benefits are independent of wage rates, these would be entirely separate items. On the other hand, if these are substitutes and both come out of the same package, workers would wish only those benefits that are valued more highly than the reduced rate at which wages grow. If the latter situation is correct, we would expect to see fringe benefits—nonmonetary returns—becoming an increasing proportion of the total wage package as three things occur: (1) as money wages increase, (2) as the income tax rate increases, and (3) as inflation occurs. Let us see why.

First, let us assume that workers value some hypothetical change in working conditions or benefits at $100 per month. That is the maximum amount workers would be willing to pay to have the change made. Thus, if a contract calling for the improvement causes monthly take-home pay to be more than $100 lower than it would have been without the change, workers would choose the higher wage over the benefit. Assume that workers would have received $125 more in gross pay in the absence of the change. With no taxation, it is obvious that the higher wage would be chosen. But if the marginal income tax rate is greater than 20 percent, workers would take home less than $100 in increased wages and would therefore lose utility by choosing the benefit. Thus, the greater the tax rate, the greater the probability a fringe benefit will be chosen over the added wages, if in fact wages and benefits are not independent.

Furthermore, as money wages rise, the marginal tax bracket of workers rises, and the cost in lost wages for fringe benefits falls. Obviously, the same conclusion can be drawn as that from increased tax rates. Benefits that would have been too costly previously, now become less costly in terms of lost wages.

Finally, of course, inflation increases the effective tax rate, because tax rates are set in terms of money income, not real income. For example, suppose all prices double; concurrently, the incomes of all individuals double also. One might think that no one would be any better or any worse off than before the increase, because each person could continue buying exactly the same things as before, but such is not the case. Doubling one's income moves one into a higher marginal tax-rate bracket, possibly from the 20-percent to the 30-percent level. Thus, any increase in income that is offset by increased prices—always the case during inflation—must make people worse off, since they have less of their income to spend. Inflation, which causes higher real tax rates, will decrease the cost of fringe benefits relative to foregone wages.

Therefore, because of the tax structure, we would expect the value to workers of fringe benefits relative to pure wage increases to rise during periods of rising wages, increasing tax rates, and inflation. This is, of course, what we have observed in the past in many, many industries. During a period of rising money wages, increasing taxation, and inflation, the percentage of the increase in the total package attributed to nonwage benefits has risen. It appears then that benefits and wages are to some extent substitutes. There is, consequently, reason to believe that some portion of the environmental regulations, particularly in-plant regulations, are at the expense of increased wages. The magnitude of the shift is an empirical question.

SUMMARY

In this chapter we have presented an analytical framework within which the benefits forthcoming from the imposition of environmental controls on mineral-extracting firms can be analyzed. As we showed in Chapters 7 and 8, such controls can impose substantial economic costs on society, particularly through reduced investment. It follows then that, in order to formulate effective public policy, one must be able to evaluate the benefits from controls in order to compare the costs and benefits.

We began by showing that the theory of benefits from controls for the mineral-extracting industries is quite different from the typical theory normally employed to consider manufacturing firms. The difference is a result of three features of the mineral extracting sector:

1. Many of those affected by the pollution are associated with the firm in other ways—i.e., they may lease land to the mine, sell intermediate inputs, or be employed by the mine.
2. Many of those damaged established their property rights after the pollution began or with full knowledge of the potential pollution.
3. The damage is generally quite localized (and may be confined to areas with low population densities).

For analytical simplicity, we separated those damaged into two groups, owners of physical capital—particularly land—and individuals. It was shown that, in both cases, the net effect of the imposition of environmental controls depends critically on the temporal conditions of ownership. For example, such controls may increase the value of an individual's residence, but if that individual had entered the area after the pollution began, he had already been compensated because he was able to purchase the residence more cheaply, and the controls would involve double compensation. Indeed, while the controls would compensate the individual who lived in the area both before and after the pollution began, they will not compensate that person who left the area as a result of the pollution and was forced to sell his property at a reduced price. Hence, measuring the benefits from controls as the increase in value may give an inaccurate picture of the total benefits, since it is necessary to consider the incidence of damage as well as the existence of pollution.

We next turned to an approach of quantifying the monetary benefits from controls. To do this we performed some conceptual experiments. Using data from Ontario, we estimated the increase in revenues that would have occurred had the existing controls been instituted earlier. We posited that the controls had been instituted in 1961 and calculated the increase in revenues had the land "disturbed" by mining been used in agriculture or forestry. The main finding was that the increase would have been minute. Hence, it would appear that if a case is to be made that the benefits of controls outweigh the costs, this case must rest on psychic or nonpecuniary returns.

Finally, we considered the effect of environmental controls on the employees of the mineral-extracting firms. The primary point made in this analysis is that, while some of the cost of compliance is paid by the firm in the form of reduced profit and a lower rate of return, and while some of the cost may be shifted forward to the consumer in the form of higher prices, the employees who benefit from the regulation will partially pay for them in the form of lower wages. Environmental legislation that improves working conditions would be expected to increase the supply of labor to that sector, and such an increase in supply will lead to a reduction in the wage rate.

NOTES

1. This situation in which all of the firms would benefit if they reduced pollution but independent action would not lead to reduced pollution is an application of what is called "the prisoner's dilemma." The problem exists that the firms impose externalities on one another. If one firm reduces its pollution and the others do not, the return to the firm is reduced; so, in the absence of collective action, the individual firms will not invest in reducing their pollution. For further discussion and an additional application, cf. O.A. Davis and A.B. Whinston, "Economics of Urban Renewal," *Law and Contemporary Problems* (Winter 1961): 105–17; see also the reference to that article in Chapter 6 of this work.

2. *Ontario Statistics, 1975*, vol. 2, Economic Series, Ontario Ministry of Treasury, Economic, and Intergovernmental Affairs, Toronto, 1976.

10

Resource Stockpiling and Conservation

Conservation of natural resources has recently become an extremely important issue. All aspects of the resource-conservation question are being hotly debated not only in academic circles but also in the popular media. In this chapter we will analyze the economic and historical feasibility of resource stockpiling and extraction holdbacks in order to assess the validity and relevance of the resource conservation arguments.[1]

Of course, the topic of natural-resource conservation is not new. Conservation has received a certain amount of attention throughout history. At the end of the nineteenth century ardent conservationists were extremely vocal and had a significant impact on public policy in the United States. Many writers thereafter, both in the popular and in the academic press, have published extensively in the area of conservation. The emphasis in most cases throughout the period has been that natural resources are being exploited far too rapidly and that government must intervene in order to save resources for future generations.

Within the past few years the interest in the conservation and exploitation of natural resources has increased astonishingly. The recently renewed interest

---

This analysis originally appeared as a monograph, Gerhard Anders, W. Philip Gramm, and S. Charles Maurice, *Does Resource Conservation Pay?* (Los Angeles: International Institute for Economic Research, 1978).

is due to several factors. First, there is the increasing emphasis on the "doomsday" theory, the idea that the world is doomed because of running out of natural resources. Such predictions have been given increased respectability by the use of the computer, which has provided a scientific facade for the same projection techniques that predicted doom centuries ago. Many conservationists emphasize that there must be control of resource extraction or future generations, lacking natural resources, will be impoverished. The Canadian government, for example, includes "conservation of mineral resources for long-term domestic requirements" as one of the elements in a mineral policy for Canada.[2] The implication of such a goal is that Canada is depleting its resources too rapidly. In other words, the government should consider delaying extraction in order to make future generations better off. In the United States the government has withheld from production its naval petroleum reserves since the administration of Theodore Roosevelt. These reserves were to protect the military in the future from "excessively high" oil prices. The Canadian government is now stockpiling uranium for the future. These are only a few examples of governmentally enforced long-term conservation of natural resources.

A second reason for the increased interest in stockpiling resources and increasing conservation is the recent Arabian oil embargo. Nations, wishing to protect themselves against the possibility of future embargos, not only of oil but also of other mineral resources, are attempting as a national policy to stockpile mineral resources. This stockpiling of resources is supposedly an efficient method of protection both from total embargos by foreign countries and from severe price increases due to international cartel price fixing. Resource stockpiling is part of a drive for self-sufficiency by the United States in key natural resources, and most nations of the Western world are currently considering similar policies.

Finally, it has been argued, with little theoretical or empirical foundation, that countries with a significant amount of the known world reserves in a natural resource—such as the Arabian oil—have the economically viable option of leaving the resource in the ground and simply letting it appreciate in value. Investment in sterile mineral resources has come to be a considered alternative to conventional investment. Thus, conservation of natural resources, extraction holdbacks, and resource stockpiling has received increased interest in the press, from governmental agencies and from private businesses.

Seldom is anything learned from history, because people tend to view each problem as unique and without precedent. In the cases of production holdbacks, stockpiling, and resource conservation, we have an excellent historical case study in the conservation movement in the United States during the early part of the twentieth century. The study of the conservation era can shed considerable light on the public policy debate regarding stockpiling and resource holdbacks as well as long-term conservation. Specifically, we can analyze empirically the costs of holding back resources from production during the twentieth century. In this study we shall attempt to measure whether long-term resource conservation has, in fact, made future generations better off economically. In such a framework the historical benefits and costs of resource conservation can be assessed.

The analysis begins with a simple measure of the cost of resource conservation. While the basic analysis involves holdbacks by private producers, we shall show that the theoretical structure applies to governmental decision making as well. Thus, the model and empirical results of this study cover both private producers and governments. Portions of this study are devoted to an empirical analysis of conservation and stockpiling in the past. Specific attention is given to the economics involved in the Teapot Dome scandal and the holding of the naval oil reserves.

The conclusions of this chapter, which follow directly from simple economic analysis and from the existing data on resource prices in the twentieth century, are largely in contradiction with the general opinion concerning conservation of depletable resources. Our conclusions are, first, that at any time during the twentieth century, enforced long-term conservation or production holdbacks of mineral resources would have been a poor economic decision, not only for the generation giving up the consumption of the conserved resources but also for the later generation that did obtain the use of the resources. Second, we conclude that whatever justifications may be used for stockpiling, such as military or political considerations, the exhaustion of resources has historically not been a valid justification.[3]

THEORY OF EXTRACTION HOLDBACKS
AND STOCKPILING

In order to examine the costs and benefits of government short-term or long-term resource holdbacks, we will begin our analysis with a study of why private firms might follow such a policy. We can then easily adapt the theory to fit governmental decision making.

A private owner or leasor of a nonrenewable resource deposit might postpone exploitation of the deposit or curtail extraction significantly for two economic reasons. First, if the firm has significant monopoly power in the relevant mineral (that is, if the firm's rate of output has a substantial effect on the world price of the resource), it may withhold extraction to increase the market price. While exploitation of monopoly power might induce a firm to hold back output, it is not the argument generally proposed in recent discussions of stockpiling, and it is certainly not an argument for long-term conservation. Neither is it the point in which we are interested here.*

The motive for a firm to withhold extraction or to reduce it substantially that is relevant for our study is the expectation that the value of the resource will appreciate sufficiently to compensate for the lost income that would have been derived from extraction. Obviously, if a firm thinks the price of a resource owned or leased by the firm will rise sufficiently, the firm should reduce or postpone extraction of the resource to take advantage of the appreciation. The crucial question is, therefore, what constitutes a sufficient price increase—that is, how large must the expected price increase be in order to make holdbacks profitable? But simply an increase in the value of the resource is not enough. The increase in the expected value over the relevant time period must exceed the returns that could be expected from selling the resource now and investing the income over the relevant time period. In making the decisions concerning the timing of extraction, the two alternative returns must be compared. Clearly, the key variable for the decision—but the variable so frequently neglected by those who have been arguing in the popular press for enforced stockpiling and conservation—is the rate of interest. Let us now examine the effect of the rate of interest on production holdback decisions.

The theory is quite simple. If someone owns a sterile asset worth $1.00 and the relevant rate of interest is 10 percent, that person should hold the asset until the next period if the price of the asset is expected to rise above $1.10†. If the price expected in the next period is less than $1.10, the asset should be sold, because the value of a dollar invested now at the market rate of interest will be $1.10 in the next period. To generalize, a business would withhold the sale of a sterile asset, including a mineral, if it expected the value of the asset to increase at a rate greater than the relevant rate of discount. Otherwise, the business would maximize wealth by selling the asset and investing the returns. This analysis holds whether we are discussing the decision to withhold an entire deposit or to increase or reduce the rate of extraction.

---

*Neither are we concerned here with the "common property" problem. This problem evolves where several parties jointly own a resource deposit. No one would have the incentive to hold extraction to the optimal level, both from a social and a private point of view, because if one firm does not extract, some other will extract more rapidly, and the resource will be used uneconomically. The reader is referred to Chapter 6 for further discussion of this problem.

†A sterile asset is defined as an asset that yields no flow of services to the holder. If an asset yields a flow of services, the decision rule would be to sell the asset if the relevant interest rate exceeds expected price appreciation plus the value of services the asset will yield over the period. Mineral deposits would under normal circumstances be sterile whereas land might yield a crop or provide recreation services to its owner.

Let us emphasize in the case of minerals that this analysis applies to the price of the resource less the cost of bringing the resource to market, rather than to the market price of the resource alone. A firm may well postpone profitable extraction even though the market price of the resource is expected to remain constant, if the cost of extracting and selling is expected to decrease sufficiently (a "sufficient" decrease being defined as a decrease at a rate absolutely greater than the rate of interest). Of course, when cost generally varies in about the same way as price, one can concentrate exclusively upon price variations when analyzing stockpiling decisions. For example, if price is expected to increase by 10 percent, so long as cost increases by 10 percent also, profit will increase by 10 percent. Therefore, when analyzing the economics of selling now or postponing sales, one is justified in looking at the variation in selling price when price and cost generally vary together.

Thus the most relevant test of the economic feasibility of resource stockpiling is a comparison of the discounted difference (at the relevant interest rate) between price and cost in some future period with the same difference in the present period. If the discounted future difference exceeds the present difference, production of the resource should be withheld. If the opposite is the case, production should occur in the present. The generalized test would be*

$$\frac{(P_t - C_t)/(1+r)^t}{(P_0 - C_0)} \quad \begin{array}{l} < 1 \text{ sell} \\ > 1 \text{ hold} \end{array}$$

where P_0 and P_t are, respectively, the price in the present period and the price expected in the target sale year t.
C_0 and C_t are respectively the cost in the present period and the cost expected in the t-th period.
r is the relevant rate of discount.

To get some idea of the rate at which the value of a resource must increase in order to make stockpiling economically feasible, consider the following simplified example. Suppose that the value of a resource at the present time, year zero, is $100, and let us assume a time horizon of 25 years. Figure 10.1 shows the way in which the $100 appreciates over the 25 years under the assumption of four different rates of interest: 3 percent, 6 percent, 8 percent,

---

*From earlier discussions, the reader can see that the numerator gives the discounted value of the net returns from delayed production, while the denominator is the net return from production now.

FIGURE 10.1

Power of Compound Interest

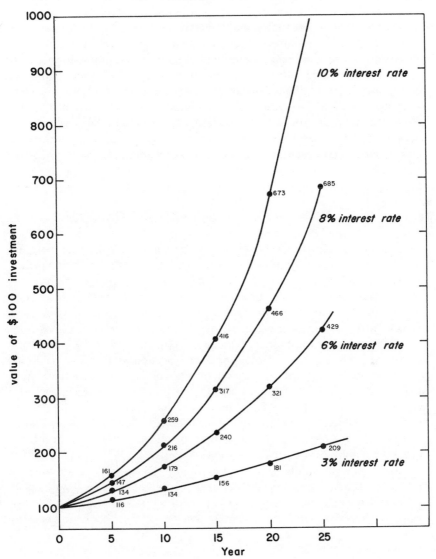

and 10 percent. In the figure the years are plotted on the horizontal axis, and the value of the investment is plotted along the vertical axis. The appreciation graphs for each rate of interest are shown. Note how the effect of an increased rate of interest begins quickly and compounds geometrically. For stockpiling to pay the value of the resource, its price minus production costs would have to increase more rapidly than the rate shown by the relevant graph, since at each of these plotted interest rates the curves show the break-even price required for stockpiling to be economically feasible. The power of compound interest is often underestimated. As an example, had the Dutch West India Company, which bought Manhattan Island from the Indians for $24 worth of beads in 1626, borrowed the money at 8-percent interest the value of the debt would be $11.98 trillion today. In fact had the relevant interest rate been 8 percent in 1626, Manhattan Island would have been a terrible long-term investment; the Indians would have been shrewd to sell the island and then invest the $24 at 8 percent.*

RESOURCE PRICES AND PRODUCTION COSTS

For empirical analysis of enforced conservation we shall, because of some data limitations, concentrate exclusively on variations in the prices of minerals. If costs and prices have generally varied proportionately, we are safe in relying only on price variations for our analysis. If, for example, both price and cost increase by 10 percent, the price of the resource in the ground rises by 10 percent also. Thus, the increase in the market price of the resource gives us an accurate estimate of the increase in the price of the resource in the ground. But if costs and prices have not varied together, variations in prices alone do not give adequate estimates of variations in the price of the resource in the ground. In this section, we use data to show that there is good reason to believe that the variation in costs, basically extraction costs, has been quite similar to the variation in mineral prices throughout the twentieth century. The tests that we carry out in this section give justification for our use of changes in mineral prices to analyze resource conservation in later sections.

Cost and value data exist for the mining industry of the Province of Ontario in the following form: expenditure upon labor (the total wages paid) per unit of ore extracted, the total amount of ore extracted, and the total gross value of ore extracted.† The data were available for most of the twentieth

---

*This illustration of course ignores the fact that a real opportunity to invest the $24 at 8 percent safely for three and a half centuries existed for neither of the parties.

†The "value" used does not represent actual market values realized, but a value calculated by applying selected commodity prices to assay contents of ores or concentrates and taking into account recovery factors. However, since this value has been calculated in a consistent manner over time, it may be used for the present purpose.

century. We wish to test whether or not the wages paid per 1,000 tons of ore hoisted has varied in about the same way as the market value of that 1,000 tons of ore over the years.

This seems to be a reasonable test of whether the cost of extraction has risen at about the same rate as the market price of the resource. If technological change has done anything, it has probably made mineral extraction less labor intensive over the century. In any case, if total costs have historically fallen relative to prices, one would think that labor costs would be a good indicator of cost changes in these directions when technological change has been relatively labor saving.[4]

There are several tests of the hypothesis that extraction costs have varied in about the same way as mineral prices that can be carried out with the existing data. First, we can compare the yearly *directions* of change in the value (that is, the yearly average price) per 1,000 tons of ore hoisted in the Province of Ontario and the total wage bill per 1,000 tons for the 72 years for which data are available, 1900–72.

Labor cost and value moved in the same direction in 60 of the 72 years, or about 83 percent of the time. In six of the 72 years, or about 8 percent of the time, the labor cost rose as the value fell. In only six years did labor cost per 1,000 tons of ore fall while the value of the ore increased. Thus, there is little evidence that costs fell as prices rose during the period.

The next test performed concerns whether the *magnitudes* of the changes in value per 1,000 tons differed significantly from the magnitudes of the changes in costs. We first calculate the yearly percentage increase or decrease in both the value per 1,000 tons and the labor cost per 1,000 tons for each year. We then can calculate the *difference* between the two percentage changes in each of the 72 years for which data are available. The mean and the standard deviation of the differences are, respectively, $X = 0.00873$ and $s = 0.1572$. Thus, the average difference in percentage changes was less than 1 percent. The calculated t-ratio was 0.437, allowing us to accept the hypothesis that there is no statistically significant difference between the percentage changes in value and in labor cost. Again, this would indicate that value and cost moved in substantially the same way.

Finally, we can compute a simple regression equation between the two variables. The regression equation is

$$\left(\frac{\text{Total Value}}{\text{Ore Hoisted}}\right) = 2.31905 + 3.3353 \left(\frac{\text{Total Wages}}{\text{Ore Hoisted}}\right).$$

The coefficient of correlation is 0.7774. The t value for the regression coefficient is 9.652. This suggests that we accept the hypothesis that the relation between labor cost and total value is statistically significant at any relevant level. The regression indicates that the value per ton rose by about $3.33 as the

TABLE 10.1

Value and Labor Cost of Ore Hoisted in the Province of Ontario

Year (1)	Value of 1,000 Tons of Ore Hoisted (Thousands) (2)	Labor Cost of 1,000 Tons (Tons of Ore ÷ Total Wage Bill) (Thousands) (3)
1900	4.21	3.27
1901	4.23	2.32
1902	5.14	1.97
1903	9.07	2.83
1904	7.47	2.64
1905	8.57	2.13
1906	13.66	3.52
1907	12.65	3.59
1908	14.28	4.34
1909	16.34	5.56
1910	16.27	4.40
1911	20.06	4.99
1912	18.82	5.13
1913	15.84	4.92
1914	10.66	3.51
1915	13.70	2.87
1916	14.57	3.04
1917	16.17	5.09
1918	20.16	4.84
1919	17.02	5.24
1920	15.81	4.96
1921	11.90	3.67
1922	14.13	3.24
1923	11.22	3.47
1924	10.95	3.19

Source: All data from *Annual Reports*, Ministry of Natural Resources Ontario, Division of Mines.

wage paid per ton increases by $1.00; in other words, the percentage of wages to value is about 30 percent. Table 10.1 presents the data used in our three tests.

These tests suggest that there has been no significant decrease in the costs of extraction relative to the price of ore in the mining industry of the Province of Ontario. To the extent that the Ontario mining industry faces technological and cost conditions similar to the mining industry as a whole, our results

Year (1)	Value of 1,000 Tons of Ore Hoisted (Thousands) (2)	Labor Cost of 1,000 Tons (Tons of Ore ÷ Total Wage Bill) (Thousands) (3)
1925	12.29	3.09
1926	10.91	2.96
1927	10.46	2.83
1928	11.82	3.03
1929	13.33	3.77
1930	13.64	3.73
1931	10.49	3.02
1932	10.07	2.80
1933	10.93	2.72
1934	13.80	2.67
1935	13.36	2.75
1936	13.28	2.84
1937	13.83	2.99
1938	12.43	2.94
1939	11.17	2.74
1940	11.39	2.60
1941	10.40	2.58
1942	9.95	2.37
1943	9.56	2.40
1944	9.01	2.43
1945	10.06	2.43
1946	8.90	2.60
1947	9.03	2.49
1948	11.26	3.25
1949	11.49	3.29

suggest that bias introduced into the measurement of the economic feasibility of stockpiling by comparing only resource prices rather than price net of cost is small. In the analysis that follows, we will use only price comparisons to gauge the economic costs or benefits from conservation. While extensive data do not exist for extraction costs, the results from our tests employing Ontario data suggest that this method of approach does not severely distort our conclusions.

TABLE 10.1 *(Continued)*

Year (1)	Value of 1,000 Tons of Ore Hoisted (Thousands) (2)	Labor Cost of 1,000 Tons (Tons of Ore ÷ Total Wage Bill) (Thousands) (3)
1950	12.62	3.44
1951	15.13	4.06
1952	13.27	4.06
1953	13.57	4.06
1954	13.90	4.11
1955	14.78	3.79
1956	15.15	3.91
1957	15.30	4.28
1958	16.27	4.43
1959	15.51	3.73
1960	15.90	3.72
1961	17.11	3.80
1962	17.51	3.98
1963	17.01	3.90
1964	16.50	3.59
1965	18.46	4.07
1966	18.47	4.48
1967	22.11	4.74
1968	18.60	3.45
1969	18.21	3.50
1970	19.66	3.76
1971	19.83	4.13
1972	22.03	4.49

CONSERVATION AS A GOVERNMENT POLICY

The relation between the rate of interest and the decision about whether to sell an asset now or hold it until later is fundamental for business decision making. Certainly, businesses are interested in making profits, and more profits are preferred to less, other things equal. One might question, however, whether such analysis applies to government, which obviously has objectives or motivations other than economic considerations. While government may have motives other than purely economic motives, the preceding economic analysis applies to governmental decision making for two extremely important reasons.

In the first place, whatever a government wishes to do and whatever the motivation, the cost to society of carrying out the decision must be considered or taken into account when making decisions. For example, increasing welfare benefits has costs to society, even though the increase in benefits is not motivated purely on economic grounds. The same thing applies to increases in the military budget or other programs carried out by governmental agencies. Even though increasing the military budget may be carried out for other than purely economic reasons, the cost of the increased budget must be considered and compared with alternative uses of the funds. Thus, any project of government involves a cost, which is the loss of the resources in alternative uses. This cost is the opportunity cost of the project.

The same type of cost consideration must also apply to natural-resource stockpiling or conservation, even though holding resources in the ground rather than extracting them involves no out-of-pocket costs to the government treasury and even though the delayed extraction is ordered from purely noneconomic motivations. No matter what the motivation, economic or otherwise, the decision to hold resources involves opportunity costs and these costs may not be inconsiderable.

If the value of the resource increases less rapidly than the rate at which the investment of the value of the resource increases, society loses economically by withholding extraction.* That is, the value of the resource must increase at a rate greater than the relevant rate of interest or there is a social cost involved. The *difference* between the increase in the value of the resource and the interest rate is the cost to society of withholding the extraction. On the other hand, if the value of the resource conserved grows at a more rapid rate than the relevant rate of interest over the time period of conservation, society is in fact better off economically for having postponed exploitation of the resource. If these circumstances occur, society benefits economically from a withholding action that may have been motivated by noneconomic considerations.

In conservation decisions, government should recognize that the true cost of withholding extraction of resources is as real a cost as the price of a good purchased, even when the motivation for conservation is noneconomic. This real cost is all too frequently neglected in the conservation literature and in discussions of the desirability of resource stockpiling.

---

*We are assuming that in the absence of the governments decision to postpone extraction private firms would choose to extract. There is, of course, no cost to forbidding something that would not have been done anyway.

The key variable in the process of analyzing the decision to delay extraction is the break-even price. The break-even price is the value in some future year that would be obtained from investing the sale price of a unit of the resource in the present year at the relevant rate of interest. For example, suppose the value of a unit of a resource today is $7.00 and the relevant rate of interest is a constant 5 percent a year. The value of $7.00 invested for nine years is

$$\$7.00(1 + 0.05)^9 = \$10.85$$

Thus, $10.85 is a "break-even" price in the following sense: Stockpiling or conservation of units of the resource would have been economically feasible if the value of the resource had risen to an amount in excess of $10.85. Otherwise, if the interest rate had been 5 percent, selling the resource at $7.00 and investing would have been preferable economically.

Even though the motive for stockpiling is for reasons entirely divorced from economic interest, the break-even price gives a good indication of the cost of stockpiling. For example, if the price in the future period happens to be $9.00 per unit, the cost is the difference between the break-even price and the actual price—in this case, $1.85—per unit stockpiled. Of course, if the price in the future period exceeds $10.85, society benefits to the extent of the difference, even though the motivation is noneconomic. In any year the ratio of the break-even price to the actual price indicates whether society does or does not benefit economically by stockpiling. The possible magnitude of these benefits and costs historically will be estimated in the statistical section of this chapter.

The second important reason for applying economic analysis to the governmental decision to withhold extraction of resources concerns long-term resource conservation. The resource-conservation movement is and has been involved basically in plans to make future generations better off economically by postponing far into the future the extraction of some natural resources. What is extracted now will not be available later; thus later generations will, it is alleged, be impoverished if present generations use up resources at the market-determined rate.

We have indicated the logical method to measure the cost or benefits of long-term conservation of resources. If the real value of the resource appreciates less rapidly over time than the relevant rate of interest, the future generation would be made better off economically if the philanthropic generation exploits the resources and invests the return at the market rate of interest. On the other hand, if increasing scarcity causes the value of the natural resource to appreciate more rapidly than the interest rate, the later generations benefit more by having inherited the resources in the ground. Again, when a society considers conservation as a method of endowing future generations, it cannot neglect the cost of the endowment; that cost is the lost opportunity of investing the return from the resources at the going rate of interest.* The endowing generation bears the cost of foregoing income from using the resources. The endowed generation bears a cost when it could have been made better off if the previous generation had exploited the resource and invested the returns in capital for the later era. Which program would leave future generations better off (leaving resources or exploiting the resources and investing) is an economic question that can be estimated empirically.

Just as an aside, the question of long-term resource conservation brings up two additional problems. First, there is the distinct possibility that technological change will make the conserved resource obsolete. In this case early generations sacrifice in order to benefit the future. But then the future generations have little or no use for the resource that the earlier group gave up. Second, there is the high probability that the earlier generation sacrifices in order to make a vastly richer generation even richer. If, for example, Americans in 1830 had stockpiled whale oil to protect the Americans of the 1970s against the loss of a lighting fuel, not only would such fuel be technologically obsolete today, but a relatively poor generation would have sacrificed to attempt to make a relatively rich generation richer.

The relative wealth of the sacrificing and benefitting generations is a relevant factor when making real world calculations of the costs and benefits of the conservation of natural resources. If one accepts the hypothesis of diminishing marginal utility of income, upon which the progressive income-tax system is based, then in order to make a comparison of the costs and benefits of resource conservation, the break-even price would have to be adjusted for the diminishing marginal utility of income that would occur if the benefitting generations were significantly wealthier than the sacrificing generations. Given the growth pattern of the United States, this is a significant factor, since the income level of the current generation is approximately 20 times the income level of the generation a century ago. While such adjustment would significantly lower the attractiveness of resource conservation, in our empirical analysis, we will neglect the relative wealth of the sacrificing and benefitting generations and concentrate solely on the efficiency of the intended transfer.

---

*It should be noted that for the current generation it does not matter whether the society in question actually invests the return from selling the resource or consumes it, because the market rate of interest would naturally be determined at a rate at which the average member of society would be indifferent between investing and consuming. This last unit of consumption that occurs in society must yield a rate of return to the consumer that is exactly equivalent to the return of an alternative investment that could have been made with the same expenditure of resources.

To summarize, the value of a natural resource in the ground in any period is the price at which the resource would sell minus the cost of extraction. If the decision criterion for conservation is to make some future generation better off than it would otherwise be, the decision to force production holdbacks would be uneconomical when the value of the conserved resources increases at a rate less rapid than the relevant rate of discount. Otherwise, as noted, exploitation and investment of the returns would maximize the wealth of the future generation. We will concern ourselves here only with the economic feasibility of conservation or stockpiling decisions regardless of motivation, in order to isolate the social costs of such decisions.

HISTORICAL ANALYSIS

Conceptually, we could determine whether enforced conservation of some minerals would be a good investment for the future, or we could at least estimate the economic cost of such conservation, even though the conservation is not done from economic motivation alone. In theory, we would predict the price of the resource in some future year. Then we would calculate the value in that future year of a unit of the resource sold now and invested at some relevant rate of interest. The cost or benefit of conservation would then be measured by subtracting the value of the investment from the projected value of the resource at the end of the conservation period.

This method is, of course, highly impractical. No one can reasonably predict the price of a natural resource or the market rate of interest a decade or century from now. One can, however, look at the historical facts concerning actual resource price growth and actual market interest rates to get some idea about what the costs of and returns to conservation have been in the past.

To this end we shall examine the historical facts concerning resource prices. We chose 14 depletable resources for which price data were available throughout most of the twentieth century. These resources are listed in column 1 of Table 10.2. The average purchase price of 11 of the resources in the United States in 1900 is given in column 2. In the case of three resources, crude petroleum, lime, and magnesium, price data were not available until later years, noted at the bottom of the table. In these cases the prices in column 2 are prices in the first year in which consistent data were available. Column 3 lists the average price of each resource in 1975.

The following types of tests are carried out. Using four rates of interest or rates of discount (the rate of return on AAA bonds, the average yearly rate of return in manufacturing before taxation, and two measures of a pure interest rate), we calculate the value in 1975 that could have been realized by investing the value of a unit of the resource in 1900 (or in the first year for which data are available). The costs or benefits of conservation or production holdbacks can

TABLE 10.2

Price in 1900, Price in 1975, and Break-Even Prices in 1975, Employing Four Interest Rates and Percentage of Break-Even Prices to 1975 Sale Prices for 14 Depletable Resources.

Resource (1)	Price 1900 (2)	1975 (3)	R = AAA		CPI + 2%		CPI + 3%		R = BTM	
			Real Break-Even Prices							
			$ or ¢ (4a)	% (4b)	$ or ¢ (5a)	% (5b)	$ or ¢ (6a)	% (6b)	$ or ¢ (× 1,000) (7a)	% (× 1,000) (7b)
Aluminum	32.72	39.80	939.90	2,361	950.20	2,387	1,941.40	4,878	1,623.40	4,079
Bauxite	3.87	15.00	111.20	741	112.40	749	230.60	1,531	192.00	1,280
Coal	1.04	18.75	29.90	159	30.20	161	61.70	329	51.60	275
Copper	0.17	0.64	4.80	745	4.80	753	9.80	1,539	8.20	1,287
Crude petroleum	0.62†	7.52	14.30	191	15.00	200	29.30	389	12.90	172
Gold	20.67	162.25	593.70	366	600.20	370	1,226.40	756	1,025.60	632
Iron ore	4.00	18.62	114.90	617	116.20	624	237.30	1,275	198.50	1,066
Lead	4.41	21.60	126.70	587	128.10	593	261.70	1,211	218.80	1,013
Lime	3.68*	22.18	89.00	401	91.00	410	179.00	807	91.20	411
Magnesium	1.81‡	0.82	22.60	2,758	19.50	2,374	33.60	4,093	3.90	481
Nickel	27.00	210.50	775.60	368	784.10	373	1,602.00	761	1,339.60	636
Silver	61.33	443.00	1,761.70	398	1,781.00	402	3,638.90	821	3,042.90	687
Tin	30.00	346.00	861.70	249	871.20	252	1,780.00	514	1,488.50	430
Zinc	4.40	39.10	126.40	323	127.80	327	261.10	668	218.30	558

*Series begins in 1904.
†Series begins in 1905.
‡Series begins in 1918.

243

NOTE

Symbol Definitions are as Follows.

H.S.
Historical Statistics of the U.S. Colonial Times to 1957; Continuation to 1963 and Revisions extension to 1962. U.S. Department of Commerce

S.A.
Statistical Abstract of the United States
U.S. Department of Commerce
Bureau of Census

M.Y.
Minerals Yearbook
Metals, Minerals and Fuels
U.S. Department of Interior
Bureau of Mines

T.N.R.C.
Trends in Natural Resource Commodities
Neal Potter and Francis T. Christy, Jr.
Johns Hopkins University Press

F.R.B.
Federal Reserve Bulletin

C.D.S.
Commodity Data Summaries, 1974
Appendix 1 to Mining and Minerals Policy
Bureau of Mines

SOURCES FOR TABLE 10.2

AAA: (corporate AAA yields [Moody's] in percent per annum): 1900–18 H.S.; 1919–75 F.R.B.

CPI: (Consumer Price Index [BLS], 1947–49 = 100): 1900–62 H.S.; 1963–75 S.A.

R$_{BTM}$: (annual rates of profit on stockholders equity in manufacturing): 1900–54 generated using average of 1955–75 period; 1955–75 Quarterly Financial Report for Manufacturing Corporations, Federal Trade Commission.

Aluminum: 1900–62 H.S.; 1963–72 S.A.; 1973–75 C.D.S. Price primary ingot in cents per pound.

Bauxite: 1900–57 T.N.R.C.; 1958–71 M.Y.; 1972–75 C.D.S. Preferred series, value at mines as shipped. In dollars per long ton. Splice factor for 1969–71 = 1.29.

Bituminous coal: 1900–62 H.S.; 1963–72 S.A.; 1973–75 C.D.S. Average dollar value per short ton, f.o.b. mine. Splice factor for 1969–72 = 0.96.

Copper: 1900–62 H.S.; 1963–72 S.A.; 1973–75 C.D.S. Wholesale price in dollars per pound.

Crude petroleum: 1900–62 H.S.; 1963–72 S.A.; 1973–75 C.D.S. Average value at well in dollars per 42-gallon barrel.

Gold: 1900–71 M.Y.; 1972 S.A.; 1973–75 C.D.S. U.S. Treasury price through March 14, 1968 and Engelhard selling quotations March 20, 1968 through 1975.

Iron ore: 1900–62 H.S.; 1963–72 S.A.; 1973–75 C.D.S. Price Mesabi, non-Bessemer, in dollars per long ton. Splice factor for 1969–72 = 0.97

Lead: 1900–62 H.S.; 1963–72 S.A.; 1973–75 C.D.S. N.Y. pig lead in cents per pound. Splice factor for 1969–72 = 0.99.

Lime: 1904–57 T.N.R.C.; 1958–71 M.Y.; 1972–75 C.D.S. Average value in dollars per short ton.

Magnesium: 1918–57 T.N.R.C.; 1958–71 M.Y.; 1973–75 C.D.S. 99.8 percent pure ingots N.Y. in dollars per pound. Splice factor for 1949–53 = 6.94.

Nickel: 1900–57 T.N.R.C.; 1958–71 M.Y.; 1973–75 C.D.S. Imports of "nickel, alloys, pigs, bars, etc.," value divided by quantity (gross) in cents per pound. Splice factor for 1969–72 = 0.91.

Silver: 1900–62 H.S.; 1963–71 M.Y.; 1972 S.A.; 1973–75 C.D.S. Average price N.Y. in cents per fine ounce.

Tin: 1900–57 T.N.R.C.; 1958–71 M.Y.; 1973–75 C.D.S. Straits tin prices in N.Y. in cents per pound. Splice factor for 1953–58 = 1.00. Splice factor 1969–72 = 0.93.

Zinc: 1900–62 H.S.; 1963–72 S.A.; 1973–75 C.D.S. N.Y. slab zinc in cents per pound. Splice factor for 1969–72 = 1.03.

be measured by subtracting the value of the investment from the price of the resource in 1975. If the 1975 price of the resource exceeds the value of the alternative investment that could have been made by selling the resource in 1900 and investing the proceeds of the sale at the going interest rate, then society clearly benefits from conservation as a method of transferring wealth from one generation to another. On the other hand, if the value of the alternative investment that could have been made by producing and selling the resource exceeded the 1975 market price, conservation of the resource in question was not the most efficient method of wealth transfer, and cost was imposed on society by the conservation decision.

During both World War I and World War II, the federal government exerted a significant amount of control over interest rates as part of a wage-price stabilization policy. Using market interest rates for these periods to measure opportunity costs severely misrepresents the actual market interest rate. In a perfectly functioning capital market, the market interest rate in any period would be the inflation rate over the next period plus some pure interest premium. The pure premium would be determined by the relative strengths of time preference and the productivity of capital. In each time period the forces of competition would drive the market rate to the sum of the inflation rate plus the pure interest premium if both savers and investors knew what the inflation rate over the period was going to be. Clearly, in any market, estimates are at variance with actual experience. It seems realistic to assume on average, however, that in the absence of governmental constraints the interest rate over long periods of time would tend to reflect both the inflation rate and some pure interest premium. Our two constructed interest rates consist of percentage change in the consumer price index plus a 2-percent interest premium and the percentage change in the consumer price index plus a 3-percent interest premium. These constructed interest rates follow a pattern similar to the market rates except during periods of government credit market controls such as World Wars I and II and periods of unanticipated inflation.

In Table 10.2 the break-even price for the 14 depletable resources is calculated for the four different interest rates. Break-even prices of Table 10.2 show the value in 1975 that would have been obtained by selling the resource in question at its market price in 1900 and investing the proceeds of that sale at the given interest rate. The formula for calculating this break-even price is

$$P_{1900} \prod_{t=1900}^{1975} (1 + r_t) = \text{breakeven price}$$

where $\qquad P_{1900} = $ The resource price in 1900 A.D.

$$\prod_{t=1900}^{1975} (1 + r_t) = \text{The product of the 75 terms } (1 + r_{1900})(1 + r_{1901})$$
$$(1 + r_{1902}) \ldots (1 + r_{1975})$$

As an example, let us take bauxite, the second resource listed in Table 10.2. Assume for simplicity a constant 4-percent rate of interest in each year from 1900 through 1975. If $3.87 had been invested in 1900 at a yearly return of 4 percent, the investment would have been worth $76.24:

$$\$3.87(1.04)^{76} = \$76.24$$

Thus, under these specialized assumptions $76.24 is the "break-even price" for bauxite. This means that the price of bauxite would have had to have exceeded $76.24 in 1975 in order for long-term conservation of bauxite to have made society in 1975 better off than it would have been had society in 1900 invested the price of bauxite in 1900, $3.87, at a 4-percent yearly rate of interest. Since the price of bauxite in 1975 was actually $15.00, the break-even price was five times the actual price. Increased conservation would not have been a good investment if the earlier generation simply wished to improve the 1975 generation economically.

The relevant question to be answered by the analysis presented in Table 10.2 is whether the growth in resource prices between 1900 and 1975 was greater than or less than the value that could have been obtained by investing the proceeds of the sale of the resource in 1900 at the AAA corporate bond rate, at our two definitions of a pure interest rate, or at the before-tax rate of return in manufacturing.* Columns 4a, 5a, 6a, and 7a show the values that would have been obtained by investing the sales price of the resource in question in 1900 at the AAA corporate bond rate, the percentage change in the consumer price index plus 2 percent, the percentage change in the consumer price index plus 3 percent, and the before-tax rate of return in manufacturing, respectively. Columns 4b, 5b, 6b and 7b show the percentages of respective yields on investment at the various interest rates relative to the purchase price in 1975.

In no case for any of the 14 depletable resources would stockpiling from 1900 through 1975 have been a viable economic alternative to simply investing the proceeds of the sale of the exploited resources in 1900 at the AAA corporate bond rate. The average value of the investment obtained by exploiting the 14 resources in 1900 and investing the proceeds of their sale in AAA corporate bonds was 733 percent of the sales price of the same resources in 1975. Of the 14 resources, in only two cases did the value of the investment fail to exceed 200 percent of the value of the resource in terms of its 1975 sale price. These two resources were coal and crude petroleum, both of which

---

*Since available data do not exist on the before-tax rate of return on manufacturing before the 1940s, the average for the years for which data are available is used as a proxy rate for earlier years. Since the lowest rate for the before-tax rate of return on manufacturing investment is large compared to our other interest rates, this proxy does not affect the conclusions of our analysis.

experienced a rapid acceleration in price between 1973 and 1975. Even so, the value of a AAA corporate bond investment of the proceeds of selling coal in 1900 was 159 percent of the sales price of coal in 1975. The value of the investment that could have been obtained in AAA corporate bonds by selling crude petroleum in 1905, the first year for which price data are available on a consistent basis, was $14.30 per barrel, which was 191 percent of the average market price of crude petroleum in 1975.

Columns 5a and 5b of Table 10.2 show the break-even prices and the ratio of the break-even price to actual 1975 price for the 14 depletable resources, using as the interest rate the percentage change in the consumer price index plus a pure interest premium of 2 percent. The results obtained are essentially the same as those shown in columns 4a and 4b, since over the 76-year period the percentage change in the consumer price index plus a 2-percent premium has roughly tracked the AAA corporate bond rate except during periods when government exerted significant control over the market interest rate. For each of the 14 resources the value of the investment that could have been obtained by selling the resources in 1900 and investing at a discount rate of the percentage change in the consumer price index plus 2 percent exceeds the market price in 1975. On average, the value of the investment that could have been obtained by selling the resource in question in 1900 and investing the proceeds of that sale at the percentage change in the consumer price index plus 2 percent is seven times the average market price for the resource in 1975.

Columns 6a and 6b show the break-even prices and the value of the break-even prices relative to the sales prices in 1975, using the percentage change in the consumer price index plus a 3-percent interest premium. It is obvious by comparing columns 5a and 5b to columns 6a and 6b that an increase of one percentage point in the interest rate has a substantial impact on the value of investment. In fact, the break-even prices from column 6a are over twice as large on average as the figures found in column 5a. The break-even prices for the 14 depletable resources as shown in column 6a are almost 13 times their average market prices for 1975.

The largest discount rate employed is the before-tax rate of return on manufacturing, which to some extent represents the investment opportunity cost of U.S. business. This extremely high discount rate produced such large numbers that it was necessary to drop the last three zeros in columns 7a and 7b. For stockpiling of crude petroleum between 1900 and 1975 to have been a viable alternative to investment at the before-tax rate of return on manufacturing, the price of crude petroleum in 1975 would have had to exceed $12,900 a barrel. The break-even price of gold would have been over $1 million an ounce. On average, the break-even price for the 14 depletable resources exceeded the market price in 1975 by 929,000 percent. Obviously, long-term stockpiling of depletable resources has not been a viable alternative to investment in U.S. manufacturing during the twentieth century.

The results of Table 10.2 cast grave doubts on the efficiency of long-term resource stockpiling for economic purposes. Clearly, during the period considered, resource stockpiling over long periods of time has not been a viable alternative to investment at the AAA corporate bond rate if the objective of stockpiling has been to increase economic welfare. When higher effective interest rates are employed, the results are even more overwhelming than those obtained with AAA corporate bond rates. This is not to say, however, that resource stockpiling with other economic motives may not have been efficient. It does, however, indicate the extremely high economic costs that have been incurred by such action. Clearly, such costs must be weighted against noneconomic advantages in order to determine the optimality of the decision. This question will be addressed in a later section with regard to the naval reserves in the United States.

SHORT-RUN STOCKPILING AND HOLDBACKS

We now turn to the analysis of short-term production holdbacks in anticipation of an increase in the price of the resource during the following one to five years, or even longer in the case of government-imposed holdbacks. We will analyze the profitability of the decision by firms to holdback extraction and the benefit or cost of governmental decisions to enforce or subsidize such holdbacks.

Certainly, firms historically have held back otherwise profitable extraction in the hope that an increase in the price of the resource would more than offset the lost income from investment of the returns. We should emphasize, however, that we are not concerned here with a firm's nonproduction or absence of exploitation of a field or reserve because production would drive marginal cost above marginal revenue. Clearly, firms do not attempt to exploit an entire area or all leased areas during one period. We are concerned, as are others who are currently discussing stockpiling, primarily with holdbacks of production that would have been otherwise profitable during a period; the holdbacks that occur because the firm expects the change in price to be so great that the profit from the same production in some future period would be larger, even when the opportunity cost of the holdbacks is considered. This type of conservation or stockpiling results from the expectation of price increases.

We must emphasize also that governments may well enforce short-term holdbacks for strictly noneconomic motives—military, self-sufficiency, and so on. We simply wish to examine the social costs of these governmental activities, just as economists would examine the cost or benefit of any action of government. These real costs of enforced holdbacks are generally ignored in most discussions of the subject.

As was the case in the section on conservation as a government policy, we shall assume that changes in resource prices indicate profits or losses from holdbacks. Thus, we have assumed that, in general, prices and extraction costs have changed in about the same way both in direction and magnitude. The analysis of the section on resource prices and production costs provides the justification for employing this simplification.

We calculate for the 14 resources listed above the number of years in which stockpiling or holdbacks for one or more years *would not have resulted in lost income,* when the opportunity cost (the lost returns) of investment is considered. We assume that stockpiling would have been economically feasible—i.e., resulted in profit between two years—if the market price in year two is greater than $(1 + r)$ times the price in year one. Otherwise, stockpiling would have been a bad investment economically.

Table 10.3 shows for each extractive industry the number of years and the percentage of the total time it would have been economically feasible for firms to have held back some production in anticipation of price increases using each of the four interest rates described above. Table 10.3 was calculated in the following way: Beginning in any year, if the break-even price in the next year is greater than the actual price in the second year, stockpiling causes losses in that period. Alternatively, if the actual price in the second year exceeds the break-even price, stockpiling or holdbacks lead to profits. Using each of the four discount rates described above, if $P_t(1 + r_t) < P_{t+1}$ (where P_t, P_{t+1} = the price of the resource in year t and year t + 1, and r_t = the relevant discount rate in year t), then stockpiling the resource from t until year t + 1 would have been a good investment. If the inequality is reversed, stockpiling would have been a bad investment economically.

In the case of government, which may not be ostensibly interested in the profit or loss from stockpiling, the resulting profit or loss simply indicates whether the decision to enforce holdbacks has an economic cost or benefit to society. If the break-even price exceeds the actual price, the magnitude of the difference shows the actual cost of the decision.

In Table 10.3, column 1 lists the 14 depletable, resources, and column 2 shows the total number of years during the period for which consistent price data are available for the resource in question. Columns 3a, 4a, 5a, and 6a show the total number of years that production holdbacks from one year to the next would have resulted in a gain that exceeded the investment yield at the going discount rates of AAA corporate bond rates, the percentage change in consumer prices plus 2 percent, the percentage change in consumer prices plus 3 percent, and the before-tax rate of return on manufacturing, respectively. Columns 3b, 4b, 5b, and 6b show the percentage of time for which production holdbacks at the various discount rates would have resulted in an economic gain.

TABLE 10.3

Number of Years Production Holdbacks in Anticipation of Price Increases Would Not Have Resulted in Losses for 14 Depletable Resources, 1900–75, Using Four Interest Rates

| | | Number of Years Holdbacks Would Not Have Resulted in Losses | | | | | | | |
| | Total Number of Years | r = AAA | | CPI + 2% | | CPI + 3% | | BTM | |
Resource (1)	(2)	Number (3a)	% (3b)	Number (4a)	% (4b)	Number (5a)	% (5b)	Number (6a)	% (6b)
Aluminum	75	21	28	22	29	20	26	6	8
Bauxite	75	18	24	23	30	21	28	5	6
Coal	75	26	34	25	33	21	28	9	12
Copper	75	38	50	34	45	33	44	14	18
Crude petroleum	70	26	37	25	35	24	34	11	15
Gold	75	7	9	13	17	12	16	5	6
Iron ore	75	25	33	28	37	24	32	9	12
Lead	75	31	41	31	41	29	38	15	20
Lime	71	18	25	18	25	14	19	4	5
Magnesium	57	14	24	13	22	13	22	3	5
Nickel	75	29	38	34	45	32	42	7	9
Silver	75	27	36	27	36	25	33	12	16
Tin	75	34	45	32	42	30	40	15	20
Zinc	75	36	48	32	42	30	40	11	14
Average		24	34	25	34	22	32	8	11

Employing the AAA corporate bond rate, in only 34 percent of the time during the period would firms have gained by holding back production from one year to the next. Obviously, if any storage costs were encountered in holding the resource, the percentage of time that gain would have been achieved by production holdbacks would have been lower. Employing the percentage change in the consumer price index plus 2 percent as a discount rate, again 34 percent of the time firms would have obtained gains and 66 percent of the time they would have incurred losses by holding back production from one year to the next. Employing the percentage change in consumer prices plus a 3-percent pure premium as a discount rate, the percentage of years during which production holdbacks from one year to the next would have resulted in economic gain drops to 32 percent. When the before-tax rate of return on manufacturing is employed as a discount rate, in only 11 percent of the time on average would production holdbacks from one year to the next have resulted in an economic advantage for the decision maker.

In only seven years did the holding of gold yield a rate of return that exceeded the rate of return available on AAA corporate bond rates. In 68 years between 1900 and 1975 the holding of gold would have resulted in an economic loss as compared to the alternative of holding a AAA corporate bond. Clearly, these results are determined in large part by the fact that the price of gold was pegged at $20.67 an ounce until 1934, at which time it jumped to $35 an ounce and remained at that price until 1968. Beginning in 1968 the price of gold rose substantially through 1975. In 38 years the price of copper rose more rapidly than the AAA corporate bond rate, indicating that in about 50 percent of the years gain could have been had in the copper industry by holding back production until the next year. When higher discount rates than the AAA corporate bond rate are employed, the percentage of time that production holdbacks would have resulted in economic gain drops substantially.

We can draw certain implications from Table 10.3 concerning the optimality of the rate of resource extraction during the twentieth century. The recent emphasis is that resources have been extracted far too rapidly by greedy entrepreneurs obsessed only with making profits from natural resources. Let us examine this thesis within the structure developed here.

First note that if firms were certain about the future, the returns from future price increases would be eliminated, because if prices were expected with certainty to increase more rapidly than the rate of interest, firms would withhold production, driving down future prices relative to current prices. If the reverse were expected, firms would increase present rates of extraction relative to future rates.

Under uncertainty with randomly distributed expectations one might expect profits from short-term holdbacks about half the time and losses about half the time, with the average returns from holdbacks being about zero. If firms would have gained from additional stockpiling or production holdbacks a very large percentage of the time, one might deduce that private firms exploited the resources too rapidly from a social point of view. If, on the other hand, short-term holdbacks would have occasioned losses in a large majority of the periods, it would appear that the resources may have been extracted too slowly.

As shown in Table 10.3 gains would have resulted from short-term holdbacks a small percentage of the time for most resources, using all rates of discount. Based upon the preceding analysis, firms may have extracted resources at too slow a rate in the twentieth century, but there is certainly no evidence that resources have been extracted too rapidly. Those who assert that a system that relies primarily on private firms uses up natural resources at a wasteful rate must defend this assertion on grounds other than historical data and economic analysis.

Figures 10.2 through 10.4 combine both the short-term and long-term analytical results, using the AAA corporate bond rate. The ratio of the break-even price to the actual price measured every five years from the beginning of the period is plotted along the vertical axis. Time is plotted along the horizontal axis. The ratios are plotted as the indicated lines in the three figures for each of the 14 resources.

A HISTORICAL CASE STUDY

No study of the economics of conservation would be complete that did not analyze the conservation movement that occurred in the United States at the turn of the century. Since the Teapot Dome affair has recently received attention, and because it was part of America's largest experiment in conservation, we will examine the real cost to society of conditions surrounding what was, until recently, perhaps the greatest government scandal in the history of the United States.[5] Most of those who have analyzed the Teapot Dome scandal have neglected the fundamental concept of the interest rate and have supported conservation. We will bring the interest rate into the analysis and show the cost of the Teapot Dome conservation effort. The motives of Secretary of the Interior Albert B. Fall are also viewed in a different light.

Initially, the naval petroleum reserves consisted of two reserves in California, totaling almost 68,000 acres, and a small area in Wyoming, Teapot Dome, totaling slightly over 9,000 acres. All were withdrawn from potential production in 1909 and were set aside as naval reserves during the period 1912 to 1915. Ostensibly, the government at the time of withdrawal wanted to be certain of an underground supply of oil, which would be tapped only if commercial reserves became scarce. Oil was to be held in the ground for the use of the government at some future time when the customary commercial

FIGURE 10.2

Ratio of Break-Even Prices to Actual Prices, AAA Bond Rate of Interest
(Aluminum, Bauxite, Coal, Copper, and Gold)

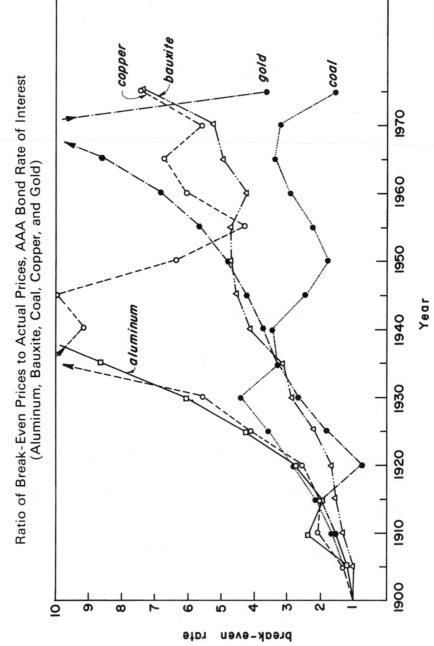

FIGURE 10.3

Ratio of Break-Even Prices to Actual prices, AAA Bond Rate of Interest
(Iron Ore, Lead, Nickel, Silver, and Tin)

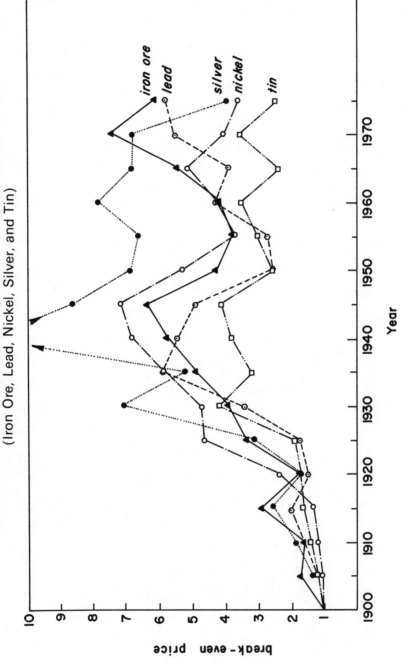

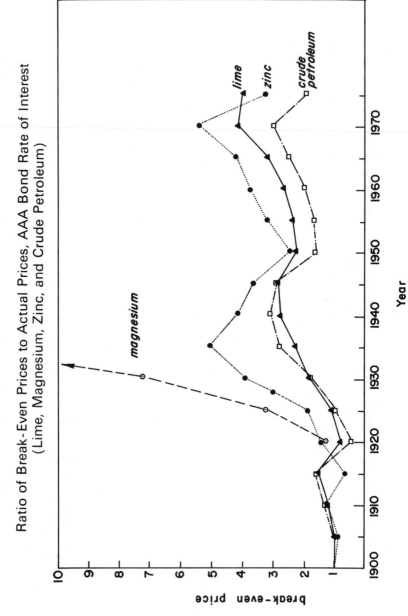

FIGURE 10.4

Ratio of Break-Even Prices to Actual Prices, AAA Bond Rate of Interest
(Lime, Magnesium, Zinc, and Crude Petroleum)

256

supplies of oil might be insufficient. Supposedly, the reserves would put the government in a good bargaining position in that "exorbitant" prices by private interests could be prevented by the threat that the navy could exploit its own reserves.

During the Harding administration, the naval reserves were transferred from the jurisdiction of the Navy Department to the Department of the Interior,[6] probably through the instigation of Albert B. Fall, Secretary of the Interior at the time. In 1922 more than half of the California reserves were leased to Edward J. Doheny, and the Teapot Dome reserve was leased to Harry Sinclair and Mamouth Oil.

At the time of transfer, an oil lease was generally negotiated with a royalty of 12.5 percent of revenues plus a bonus for the right to drill. There was no bonus in the case of the naval reserves—and this point later drew considerable attention. However, many writers during the period of investigation felt that the very complicated contract between the government and the involved oil interests called for substantially *more* than 12.5 percent royalties.[7] The contracts were complicated by requiring the building of storage tanks and pipelines for the navy and for payment in oil and other goods in kind. In this way, little money that necessarily would have been returned to the Treasury was involved. Some analysts estimated total royalties at over 30 percent.[8]

Later, beginning with the early years of the Coolidge administration, many of the oil transactions began coming to light. Public outcry in addition to the political outcry of the opposition became increasingly hostile. It began to appear that some money had been paid by the oil men to Secretary Fall to induce him to lease some of the reserves, and, of course, Fall was later convicted for bribery.

Fall's primary defense was his consistent opposition of public ownership of natural resources on philosophical grounds and in the interest of efficiency during the entire period in which he held public office. Fall's defenders argued also that privately owned wells on adjacent leases were draining substantial oil from the naval reserves.[9] Those involved in the leasing claimed that if the naval reserves were not exploited, much of the reserves would have been lost, causing a vast diminution in their value to the navy and to the government. The truth of this defense has never been established.

The government sued successfully for return of most of the naval reserves, and a few participants were convicted of bribery and sent to prison. Some of the California reserves were exploited during World War II, but much is still intact. The entire affair dropped from sight until recently, when it was compared to Watergate.

Because the corpse has been exhumed, it now might be well to examine the economics of the transaction, particularly in view of recent interest in stockpiling and in production holdbacks of oil. The implications of bribery will be ignored, even though evidence is strong that bribery existed. It will also be assumed that Fall thought, as history indicates he probably did, that resource reserves should be in private rather than in public hands. And we will also ignore the distinct possibility that considerable leakage from public reserves would have been caused by private drilling on adjacent lands. In this way attention can be focused totally upon the pure economics of the conservation issues.

During World War II, some oil was pumped from the reserves for the navy. Historians who later described the Teapot Dome affair justified the return of the reserves to the government by pointing out that oil in the reserves would not have been available had they remained in private hands. Note, however, that the whole point of the reserves was to protect the navy from shortages and high prices in time of need.

During the war period, 1942–45, crude varied in price from $1.19 to $1.22 a barrel (admittedly a regulated price). In 1922, when Fall sold the leases, oil was selling for $1.60 a barrel. Deflating the average 1942–45 crude-petroleum price by the wholesale price index to convert to 1922 dollars, crude was selling at an average $1.18 per barrel during World War II. Assuming that the royalty rate (12.5 percent) times the price represented the value of the oil in the ground, the value of the naval reserves in 1943 was $0.15 a barrel. If the reserves had been sold in 1922 and the royalties had been invested at 3 percent, certainly an unrealistically low rate of return, value of the investment in 1943 would have been $0.37 a barrel leased. At a 6-percent or 8-percent rate of return on investment, the value of the investment rises to $0.67 or $1.01 respectively for each barrel leased, a significant increase over the realized value of $0.15.

As a matter of fact, even during the Korean War, the real price of oil (the price net of the rate of inflation since 1922) was slightly *lower* than at the time of Fall's leasing. In 1950 price of crude was $2.51 per barrel. But in 1922 dollars, according to changes in the wholesale price index, the real price was only $1.59 per barrel. Assuming that royalty rate (12.5 percent) times price ($2.51) represented value of oil in the ground, the value of the naval reserves in 1950 was 30 cents (less than $0.20 a barrel in 1922 dollars). If the reserves had been leased in 1922, and royalties invested at 3 percent, the investment in 1950 would have been worth $0.46 a barrel. At 6 percent and 8 percent, the value grows to $1.02 and $1.72, respectively.

The important point, however, is to consider whether we are better off now, facing substantially higher oil prices, having held the oil in the ground rather than leasing the oil rights at even the lowest royalty rate in 1922. Also, given that the government did not lease the reserves in 1922, we may calculate the probable cost or benefit to society of this policy of conservation.

In 1974, the price of oil rose sharply due to the Arab embargo and the OPEC cartel pricing collusion. Because of this drastic jump in price from $3.66 a barrel in 1973 to an average price of $7.52 in 1975, the attractiveness of the naval reserves as an investment would appear to have improved substantially.

The recent price rise shows up more strongly in the marginal cost of crude, which grew to $10.00 by 1974.

Assuming once more that the value of oil in the ground is approximately the royalty rate times produced price, the naval reserves are worth $1.25 per barrel at a market price of $10.00. When the 1974 price is deflated to 1922 dollars, it falls to $3.41, and the royalty value falls to $0.42.

If the reserves had been leased in 1922, and royalties had been invested at 3 percent, 6 percent, or 8 percent, value of these investments would have been $0.96, $4.39, or $11.82, respectively, in 1975. Astonishingly, if the average rate of return in manufacturing before taxes is used as a social opportunity cost, the value of the investment of the royalties in 1922 would have been a staggering $2,009.97 per barrel in 1975.

The rapid increase in price in 1974 made the naval reserves a good investment, only if one does not correct for inflation and uses a pure discount rate of less than 3.5 percent. If the 1974 price is deflated, a discount rate of less than 1.5 percent is required to make the naval reserves an economically good investment. Both of these rates are significantly below the AAA corporate rate in 1922, which was 5.1 percent. The $0.20 royalty in 1922 invested in AAA corporate bonds in 1922 would have been worth $2.93 in 1975, as compared to the $1.25 value of the oil in the ground.

Using data on oil pumped out of the naval reserves since 1922 and FEA estimates on proven reserves in Teapot Dome, Buena Vista, and Elk Hills, the reserves in 1922 probably were around 1 billion barrels at the 1922 price and technology.* Assuming the reserves of 1 billion barrels were extracted in 1922, and the royalties invested in projects yielding 3 percent, 4 percent, 6 percent, or 8 percent, the real loss to the American public of holding the oil in the ground relative to leaving it for production at the average domestic price in 1975 has been $50 million, $720 million, $3,710 million, or $11,820 million.

Writing in *The Nation*, November 21, 1923, a popular political writer at the time of Teapot Dome, William Hard, castigated Fall for selling the naval leases. He concluded by saying, "Either this result [of the Teapot leases] is an absurdity or else President Roosevelt, with his policy of Navy oil conservation, was an idiot." Either might be the case, but the verdict is not quite so obvious as that columnist appeared to believe at the time. Not only have the naval reserves been a questionable economic investment from 1922 until today, but at a reasonable discount rate and rate of growth in crude prices the American people are probably losing money every day the reserves are not produced.

---

*This estimate is based upon information from several sources including personnel at the Federal Energy Administration.

Clearly, these alternative costs must be weighed against the possible military advantages of holding the reserves, but they cannot be neglected in making rational policy decisions.

EPILOGUE

From the birth of the ancient Greek science and philosophy until this century, humanity has viewed the world as possessing a given "stockpile" of resources. With a fixed environment, mankind was assumed to possess only the power to adapt to nature. The fixed resource base employed in conjunction with a growing population yielded dire prospects for mankind.

In every age, people have employed the stockpile view of resources to predict doom. All such predictions, from the Malthusian overpopulation projections to the prophecies of ecological collapse, have had basic elements in common: (1) They all assume that mankind possesses only the capacity to adapt to a fixed environment; (2) they all assume that technology is bounded by a fixed resource base; and most importantly, (3) all of them have proved to be wrong.

The world is moved not by the philosophers but by the doers, and practical men employing ingenuity and common sense have never been bound by the constraints that bind the philosopher. What history has documented thousands of times, modern science has finally recognized. In a fundamental sense, natural resources are not fixed; they are function of price, which is the driving force of science and technology. As science and technology progress, new resources come into existence and old resources become valueless. To those who walked naked in the forest, the only mineral resource was a sharp stone. By using resources that were to such a man valueless, we were able to walk on the moon. Resources are created by mankind in the same way that people create anything—i.e., by rearranging and utilizing more efficiently what is found in nature. Mankind creates its own environment and its own resources and is bounded only by the limits of imagination and freedom of action.

A case of the ability to cope with resource constraints is found in the whale-oil crisis of the nineteenth century.[10] From roughly 900 A.D. until the 1860s whale oil was an important source of lighting and lubrication. Whales were hunted in the bays and inlets of Western Europe on a sporadic basis until the middle of the sixteenth century. By the middle of the sixteenth century, the Norweigians, French, English, and Biscayans were engaged in large-scale whaling operations. As a result of such intensive whaling, the once-plentiful whales of the European coast virtually disappeared. Whalers began pursuing the whale over all the world's oceans. As the whales virtually disappeared around the coast of Western Europe, whalers sailed out to the open sea and as far north in the Atlantic as the ice would allow. By the end of the sixteenth century Europeans were whaling off the coast of Newfoundland, Iceland, and Greenland.

When the whales began to disappear in the north, the fleets turned southward and exploited the American coast. The abundance of whales along the New England coast was in part the reason the English chose to colonize Cape Cod, and for several years whaling in America was practised within the sight of land. By the 1770s America dominated the whaling industry and sent ships down the coast of the Americas. In 1791 American whalers sailed into the Pacific, and by 1820 they were taking whales off the coast of Japan. After the Indian Ocean and the South Pacific were exploited, the fleets turned North. In 1848 American whalers entered the Arctic Ocean.

The long journey from the coast of France to the Bering Straits is a testament to the ability of mankind to exploit resources and to adapt to nature. We did not run out of whale oil. Its price simply got so high that the same genius that had so augmented its supply was employed to develop its replacement. From the Arctic Ocean we turned to Pennsylvania in search of a cheaper and more abundant source of lighting and lubrication and found a great energy source in the process. A nonresource, crude petroleum, was thus made a resource by science and technology, and we adapted nature to our requirements.

We are experiencing the second major energy crisis in U.S. history. The people of the nineteenth century did not need computers to project that the supply of whales could not keep pace with the rapid expansion in demand. Sperm oil rose from $0.43 a gallon in 1823 to $2.55 a gallon in 1866. Whale oil rose from a low of $0.23 a gallon in 1823 to $1.45 a gallon in 1865. In terms of 1820s dollars, sperm oil rose from $0.44 a gallon in 1823 to a high of $1.78 per gallon in 1855.[11] In a space of 32 years, the real price of sperm oil in terms of 1820s dollars more than quadrupled. As prices rose, coal gasification became an economically feasible substitute, and there was a leveling off of the quantity of whale oil demanded, especially in Europe.

It is interesting to note that even during the height of the whale-oil crisis of the nineteenth century (1823–66), the stockpiling of whale oil was not profitable at a discount rate of 4.27 percent or higher. Only at a discount rate of 4.15 percent or lower was stockpiling sperm oil in 1823 to sell in 1866 a good investment.

By 1850 the real price of sperm oil was $1.51 a gallon, more than three times its price 27 years earlier. Such drastic increases in prices were attributed not only to the increasing scarcity of whales, which necessitated longer and more dangerous voyages, but to other sources as well. Whaling was an uncertain business requiring huge capital investment, and therefore necessitating a return to risk. In 1849 gold was discovered in California; it became common practice to sign on as a whaler and jump ship to get free transportation to California from New England. Additionally, cotton milling in New England enticed labor away from whaling and drove up the cost of labor.

In 1858 petroleum was discovered; this event in a few years ended the whale-oil 'crisis' forever. The first oil well in Lambton County, Ontario, closely followed in 1859 by the discovery in Titusville, Pennsylvania, marked the beginning of the end for whale oil as a lighting source. In the meantime, the demands of the Civil War caused whale-oil prices to boom. In addition to the increased demand, the war disrupted production. The conscription of whalers as freight ships and the capture or destruction of ships by Southern privateers caused a decline of over 50 percent in the number of ships in whaling and a 60-percent decline in tonnage. By 1866 sperm oil had reached a high of $2.55 a gallon, although its real price had already begun to fall because of competition from kerosene.

The high prices for sperm and whale oil between 1859 and 1867 provided a growing profit incentive for developing an efficient refining process for crude petroleum. Subsequent investment in research and development resulted in the production of kerosene. By 1863, 300 firms were refining petroleum products, and kerosene quickly broke the sperm- and whale-oil market, causing prices to tumble. By 1896 sperm oil was cheaper than it had been in any recorded period in U.S. history: $0.40 a gallon. But by that time whale-oil lamps had been discarded; they were no more than relics for succeeding generations.

The encroachment of petroleum into the domain formerly dominated by whale oil did not end with the use of kerosene as a lighting source. Soon, lubricants were derived from petroleum residuals, and paraffin robbed whale oil of even its ornamental uses. Whale bone and secretions that had been considered waste products in 1800 saved the industry from total extinction.

The whale-oil crisis is a case study of how the free-market system solves a scarcity problem and circumvents resource depletion. When demand increased, the price of whale oil rose and higher prices increased the number of feasible substitutes. The figures on sperm- and whale-oil production themselves are not reliable, but the rise in prices from 1820 to 1847 saw tonnage of whaling vessels rise almost 600 percent, and technological improvements in the whaling industry were numerous. In short, the rising prices caused output to increase perhaps by 1,000 percent or more.[12] Had government possessed the power and volition to attempt to ration sperm and whale oil in order to hold its price down or to levy a tax on whale oil to reap the gains from the price rise, the shortage could have been catastrophic, and the advent of kerosene and other petroleum products might have been delayed for decades. Had sperm oil been stockpiled in 1823 for use in 1900, the loss to society at a 3-percent discount rate would have been $3.79 a gallon.

The profit incentive produced by higher prices for whale oil gave an impetus to seek out and perfect alternative energy sources. The end product of this process of discovery and innovation is the Petroleum Age in which we live.

We owe the benefits and comforts of this age to free enterprise and the scarcity of whales.

The history of our first "energy crisis" and hundreds of thousands of other "crises" teach us that there is no reason to believe that we face long-term doom. If technology were suddenly frozen, some of the dire projections being made now might be realized in several hundred years or less, depending on which "expert" of the week one believes. But technology is *not* frozen, it is instead progressing at a rate unprecedented in history. The Petroleum Age will pass as did the Stone Age. The real danger is that we may foolishly restrict the exploitation of current resources, forego investment opportunities, and allow conserved resources to become obsolete. Only if we eliminate the market incentives for innovation and investment must we face a real, long-term "resource crisis."[13]

NOTES

1. *Stockpiling* is a method of conserving resources after extraction costs have been incurred. *Extraction holdbacks* refers to the conservation of resources by withholding their extraction, thereby indefinitely delaying extraction costs. While the former is normally employed to serve long-run objectives and the latter to serve short-run objectives, the distinction is simply between alternative means to a common end. It should also be noted that commodity agreements have the same effect since they involve buffer stockpiles. This point was clarified in Gerhard Anders, "The Mineral Conservation Issue—Problem or Symptom?" presented at the 80th annual meeting of the Canadian Institute of Mining and Metallurgy, April 1978.

2. See *Towards a Mineral Policy for Canada*, published under the authority of Federal and Provincial Ministers, Ottawa, 1974.

3. Others have argued that exhaustion probably does not pose a significant problem. See, for example, Richard L. Gordon, "A Reinterpretation of the Pure Theory of Exhaustion," *The Journal of Political Economy* (June 1967): 274–86; Orris C. Herfindahl, "Depletion and Economic Theory," in *Extraction Resources and Taxation*, Mason Gaffney, ed. (Madison: University of Wisconsin Press, 1967). Harold J. Barnett and Chandler Morse argued that, if anything, mineral scarcity has declined: *Scarcity and Growth* (Baltimore: The Johns Hopkins University Press, 1962).

4. During the twentieth century in Ontario annual tons of ore hoisted has consistently grown at a more rapid rate than the average annual employment in mining. These data suggest the presence of labor-saving technological development. See G. Anders, W.P. Gramm, and S.C. Maurice, *The Impact of Taxation and Environmental Controls on the Ontario Mining Industry* (Toronto: Ontario Ministry of Natural Resources, 1975). Further, the results presented in Chapter 4 would support this contention.

5. Much of the material in this section was published in a somewhat different form in W.P. Gramm and S.C. Maurice, "The *Real* Teapot Dome Scandal . . . Crude Stockpiling: A Prime Example of Poor Economics," *World Oil* (November 1975): 71–73.

6. Much of the description of the leasing arrangement is based upon two articles written at the time by Stanley Frost, "Yes—But What Are the Facts?" *The Outlook* (March 24, 1924; April 2, 1924).

7. *Ibid.*

8. *Ibid.*

9. *Ibid.*

10. Some of the descriptive analysis of the whale-oil crisis was published in W.P. Gramm, "The Energy Crisis in Perspective," *Wall Street Journal* (Friday, November 30, 1973).

11. Walter S. Tower, *A History of the American Whale Fishery* (Philadelphia: Winston, 1907), 128–29.

12. *Ibid.*, pp. 121–27.

13. This point is expanded upon in W.P. Gramm, "Debunking Doomsday," *Reason* (August 1978), 24–25, 38.

11

The Impact of
the Percentage Depletion Allowance

Thus far we have examined theoretically and empirically the impact of changes in the tax structure and in the environmental constraints on several aspects of the mineral-extraction industries. In this final chapter we examine an extremely important issue in the economics of mineral extraction: the effect of the percentage depletion allowance for tax purposes.

Economists have engaged in considerable speculation and have theorized extensively about the economic effect of a percentage depletion allowance for minerals, particularly in the oil and gas industry. The percentage depletion allowance permits a resource-extracting firm to deduct from taxes a percentage of its gross sales revenue, up to a set percentage of revenue net of costs. The majority of the economic analyses of this favorable tax treatment have been strictly theoretical in nature, with less attention directed toward empirical estimation of the actual impact of percentage depletion on the price of the mineral commodity, the quantity supplied, and so forth. A primary reason economists have generally eschewed the attempt to estimate the effects statistically is the monumental problem of obtaining relevant and accurate data.

The data problems notwithstanding, our primary purpose in this chapter is to provide such estimates. Using an approach similar to that employed in the estimation of the impact of taxation and environmental controls, we will estimate the hypothetical magnitudes of the important variables if percentage depletion had not been in effect in the past. By comparing these hypothetical estimates with the actual data, we can then obtain an estimate of the magnitude of the impact of percentage depletion.

We begin the chapter with a brief summary of the existing theoretical analyses of percentage depletion. In this literature, the major issue is the impact of percentage depletion on the allocation of resources rather than any question of equity, We next provide a simple model that may be used to predict the magnitude of the impact. This model, similar to that employed in our investigation of the effect of environmental controls and taxation, is based on the fact that the tax treatment under percentage depletion allowances shifts the marginal return on investment function and thereby alters the level of investment in the industry. We then turn to the empirical analysis itself. In this chapter we use data from the petroleum industry; however, it should be stressed that the same techniques could be used to examine any other mineral commodity. In the empirical section we first provide evidence that the petroleum industry behaves competitively and that the percentage depletion allowance operates through reallocation of investment. We then provide simulations of what output and price would have been over the period 1951–71 if percentage depletion had not been in effect. By comparing these simulated values with the actual data, we are able to obtain estimates of the magnitude of the impact.

ECONOMISTS LOOK AT PERCENTAGE DEPLETION

Economists writing about percentage depletion have been almost solely concerned with the impact upon the allocation of resources—the allocation over time and the allocation among different industries.* They have attempted to deduce theoretically whether percentage-depletion allowances attract resources (factors of production) away from sectors in which a depletion allowance does not apply to sectors in which it does. They have concerned themselves with the possible effect of percentage depletion upon changes in relative prices and have theorized about whether percentage depletion changes the rate of extraction over time.

To use economic terminology, economists have been primarily concerned about whether percentage depletion allowances are neutral or nonneutral, and if the latter, how much so. Before going on, we should define *neutrality*. A tax or a tax allowance is said to be neutral if that tax or allowance has no effect upon the allocation of resources or relative prices. On the other hand, if a tax or allowance does effect resource allocation and/or prices, that tax or allowance is not neutral. That is, if a tax is passed on or shifted to consumers or to owners of

---

*This concern does not address the question of equity or "justness," on which most of the political controversy is based. However, such questions cannot be considered if we are to stay in the realm of positive (i.e., nonsubjective) economics; they must be left in the political arena.

factors of production in the form of price changes, the tax is not neutral.

In 1955 Arnold C. Harberger published a paper that has become the classic purely economic work on the question of percentage depletion allowances in the oil and gas industries.[1] Harberger, from his testimony before Congress clearly an avowed opponent of percentage depletion allowance for oil, attempted to show that percentage depletion increases investment in affected industries by an amount somewhere between 36 and 95 percent. He reasoned that firms under competitive conditions will invest in an asset until the price of the asset equals the discounted value of the income stream net of taxation. Abstracting from large differences in risk, the after-tax rate of return on investment under competition will be approximately equal in all industries in the long run. If there are two industries and one receives a more favorable tax treatment than the other, equal after-tax rates of return call for investment in the favored industry to be carried out to such an extent that the before-tax return is lower in the favored industry, even though the after-tax returns are equal. As noted above, Harberger estimated the "overinvestment" in oil to be between 36 and 95 percent. We will set forth in the next section the mathematical model from which these results were obtained.

In 1961 Stephen L. McDonald published a paper that disagreed somewhat with Harberger's analysis.[2] He began by admitting that percentage depletion *per se* is not neutral in its effects. But, he said, the corporate income tax is not neutral either, and percentage depletion is simply one way to counteract the distortions of the income tax. Thus, the final result is not excessive investment in minerals.

He pointed out in the analysis that many authorities believe a corporate income tax is shifted in the long run in the form of higher prices to consumers. After the shifting, the industries receive a normal rate of return on investment after taxes. Normal rates of return and normal rates of capital turnover differ among industries depending upon the degree of risk and type of technology. Restoration of normal rates of return, which differ among industries, leads to a change in relative prices after the imposition of the corporate income tax. Thus, the allocation of resources is changed, and the tax is not neutral. Under these assumptions, neutrality does not require equal tax rates but discriminatory tax rates.

McDonald used some numerical examples to show that the higher the relative ratios of normal rates of return to normal capital turnover, the greater the price increase necessary to restore the normal rate of return after an equal corporate tax. Thus, these higher return, lower-turnover industries find resources being drawn away. Neutrality, therefore, requires a lower tax rate in these industries. McDonald asserted that a deduction from gross income for more disadvantaged industries—relative to industries with a lower normal rate of return and greater capital turnover—could lead to neutrality when corporate tax rates are equal. Using data from the oil and gas industry and from manufacturing in the United States, he estimated that the deduction for oil and gas necessary to establish neutrality is about 22 percent, which is his estimate of the effective depletion rate at that time.

Needless to say, McDonald's paper caused considerable controversy.[3] Most of those who disagreed with McDonald questioned his statistical results rather than the theoretical possibility of percentage depletion acting as a partial offset to the bias of a corporate income tax. Thus, it is fairly clear that in all the analysis set forth so far, a percentage depletion allowance *in and of itself* biases investment toward the affected industry. This is not to say that this bias may to a greater or lesser extent merely offset a bias in the opposite direction caused by another type of taxation. We merely say that, other things remaining the same, economists have generally agreed that percentage depletion adds to present resource extraction. How much it adds is an empirical question to which we shall address ourselves below.

All economists who have written in this area are not totally in agreement with the above assertion. Paul Davidson in a widely cited paper dealing with the oil and gas industry took exception to the generally accepted bias of a depletion allowance.[4] He argued that percentage depletion in the case of oil simply increases the value of mineral holdings.

He began by noting that the effect of favorable tax treatment in one area depends upon how responsive the supply of the resource is to changes in price. If the resource could be produced at constant costs, there would be an expansion of output and a decline in price until the after-tax rates of return were equalized at the original level in all industries. If the resource had an extremely inelastic or fixed supply, its price would be bid up, and only the original owners of the resource would benefit. No resources would be reallocated among industries. An infinite number of circumstances fall between the two extremes.

Thus, the effect of percentage depletion depends upon the responsiveness of the supply of mineral lands, that is, the relative adaptability of mineral supply to price. Davidson attempted to establish that the inducement of percentage depletion to additional or marginal exploration and development, which would not have occurred in its absence, has been rather small in the case of oil and gas. He did not analyze other minerals. He began by noting that lease bonuses and royalties are negotiable and indicated that favorable tax treatment raises the value of a landowner's minerals. If the land has no alternative use, the lease bonus and royalty are pure rent. A less favorable tax treatment would lower the value of the mineral rights and lower the bonus and royalty. There would then be no change in incentives to explore nondeveloped land, regardless of the amount of the depletion allowance, because any favorable tax treatment goes to the landowner. Only the distribution of income between producers and landowners is altered.

Davidson did state that there would be short-run effects. Owners of properties presently producing cannot renegotiate. Any alteration in the tax treatment would hurt or favor the producers, depending upon the direction of the tax alteration. A change in the tax laws unfavorable to producers would be shifted forward in some measure to consumers in the form of higher prices, the total shift depending upon the elasticities of supply and demand. Davidson argued that elasticities in the case of oil in the United States are such that elimination of percentage depletion would not result in much forward shifting.

Others took issue with Davidson's thesis. They argued that the supply of oil lands is far more elastic or responsive to profits than Davidson speculated.[5] Noting that Davidson's arguments depended in very large measure upon the supply of oil lands being almost perfectly inelastic or perfectly fixed, the critics pointed out that potential (undeveloped) oil lands vary greatly in expected yield. Areas can be ranked according to potential as to probability and cost. Under uncertainty the rankings change. All landowners within the intensive margin of exploration—expected profitability—receive rents; others do not. If the expected return rises (possibly because of a depletion allowance), the marginal lands are extended in number, and rents on inframarginal lands will increase where possible. The conclusion that a depletion allowance, by merely raising rents, would not increase exploration is correct only if the supply of new reserves is virtually fixed.

Other critics agreed with Davidson that additional tax credits do increase rent, but they also stressed the effect upon increased supply, which is to some extent offsetting to the increase in rents. The increase in supply depends on the elasticity of supply of oil, which is determined by the continuity and slope of land-quality graduations. If there are available lands of only slightly poorer quality, land rents will increase only a little. Thus, if the slope at the margin is gentle, the depletion allowance would significantly increase exploration and development.

To summarize, economists in general agree that percentage depletion causes increased investment in the affected industries. This by no means indicates that economists have consistently favored a depletion allowance in certain mineral industries; such is not the case. Economists have generally argued simply that resources are used more rapidly than in the absence of depletion.

A SIMPLE PREDICTIVE MODEL OF THE IMPACT OF THE PERCENTAGE-DEPLETION ALLOWANCE

As in the preceding chapters, we begin with the present value (or net worth) of the property. Since other forms of taxation, such as specific severance

taxes, and environmental controls are not at issue here, the present value
function developed in Chapters 7 and 8 becomes

$$PV = \sum_{t=0}^{H} \left(\frac{1}{1+r}\right)^t \{Y_t - k(Y_t - pR_t - \gamma_t C)\}$$

Reviewing the notation,

  H is the time horizon over which the firm makes its plans.
  r is the rate used to discount future profits.
  Y_t is the net return in period t. It is defined as $Y_t = R_t - V_t - F_t$, where
       R_t is sales revenue after royalty, V_t is variable cost, and F_t is fixed cost.
  k is the rate of taxation.
  p is the rate of allowed percentage depletion.
  γ_t is the allowed depreciation on investment in t-th period.
  C is the cost of the property (investment).

As was the case with the other forms of taxation, the effect of a change in
the percentage depletion allowance comes about through changes in the
marginal return on investment.

As we have done several times before for other changes, we can show
graphically the effect of changes in the depletion allowance on the rate of
investment. We repeat the analysis here for those who may have omitted the
other discussions.

In Figure 11.1 assume that MRI is derived under the assumption of a
positive percentage-depletion allowance in the extracting industry. Assume
also that the prevailing rate of return on investment for the economy as a
whole is r. So long as added investments yield an income stream with a return
greater than r, these investments will be undertaken. No investment yielding a
known after-tax rate of return less than r would be undertaken, because the
capital would be more productive elsewhere. Clearly, with a depletion
allowance and a rate of return r, the industry would invest I_0 in additional
capital.

Next, eliminate the depletion allowance. Obviously the after-tax—but not
the before-tax—rate of return for each additional potential investment falls. In
Figure 11.1 the schedule indicating the marginal return on investment declines
to MRI'. There is no reason for the rate of return in the economy as a whole to
fall very much, if at all. Thus, at the rate r, investment will be carried out to
equalize after-tax returns; this new investment level is given by I_1. There is
clearly less investment, and future output will fall. After-tax returns will remain
equal in the long run. With depletion the before-tax return must have been
below that in the economy as a whole. After the elimination, both before- and
after-tax returns will be equal. These investments will be undertaken even if

FIGURE 11.1

The Effect of Changes in the Depletion Allowance on the Rate of Investment

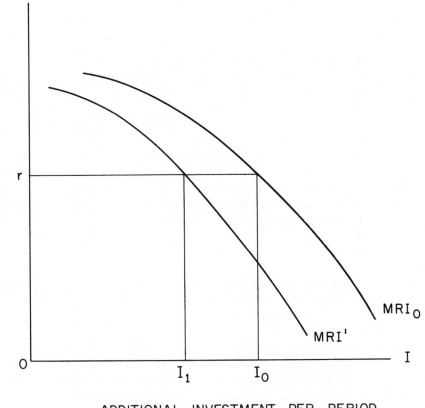

ADDITIONAL INVESTMENT PER PERIOD

land costs are bid up to the point at which profits are competed away. Thus, according to our model, a depletion allowance increases investment and exploration even if the sole beneficiaries are landowners who receive rent.

We might also note that an increase in the rate of depletion at first increases net returns after taxes for any rate of production. If profits can be increased in every period, firms will choose to increase current profits at the expense of future profits, because the present value of a unit of current profit is higher; extraction will be moved toward the present as the rate of depletion is increased, and vice versa.

Thus, in theory, a positive depletion allowance should increase investment and output above the rate that would result from a zero depletion allowance, and it should concomitantly lower prices. The real question is, *how much has percentage depletion increased output, and how much has it lowered price?* If we had perfect data from the past, we could easily estimate these results with the methods discussed here. Obviously, data do not exist for every property over the past 50 years. Therefore, we shall use a more abstract approach.

To simplify our analysis and enable us to come to grips with the problem of estimation, we will begin with the model developed by Harberger in the paper discussed in the preceding section. Assuming that Y is the discounted value of the before-tax income stream expected from an asset and W_1 the discounted value of the associated tax payments, let k be the rate of taxation, d the relevant rate of depreciation, and C the price paid for the asset. As discussed previously, competitive conditions would cause firms to pay a price for the asset equal to the discounted value of the income stream net of taxation. Since the price of the asset to the firm is the total dollar amount that can be written off as depreciation, the cost to the firm is dC rather than simply C. In the case of normal depreciation in equilibrium,

$$W_1 = k(Y - dC_1).$$

Next, Harberger assumed that percentage depletion is relevant and that p is the fraction that depletion allowances bear to *net income before taxes.* In this case for the same stream of income Y,

$$W_2 = k(Y - pY) = tY(1 - p),$$

where W_2 is the discounted value of tax payments. Harberger noted that p was not the then statutory $27 1/2$ percent that applied to gross, not net, income. Rather, he said, p was between $27 1/2$ and 50 percent, the maximum allowable depletion based upon net income.

Harberger then estimated the impact of percentage depletion versus ordinary depreciation. He again assumed that business investment is carried out to the extent that the present value of the net-of-tax stream of income expected from an asset equals the cost of the asset. The cost of an asset yielding a discounted income stream of Y and subject to typical depreciation is

$$C_1 = \frac{Y(1-k)}{(1-kd)},$$

where k and d are defined as above. Additionally, in the case of mineral extraction, exploration and development will be undertaken until receipts net of taxation equal the present value of costs, C_2. Receipts, R_2, equal $Y(1-k+kp)$, where t, p, and Y are as above. Harberger assumed from observation that tax offsets apply to 80 percent of costs. Thus, costs net of tax offsets in this model are $C_2(1-0.8k)$. In competitive equilibrium, with percentage depletion,

$$C_2 = \frac{Y(1-k+kp)}{(1-0.8k)}$$

Taking the ratio C_2/C_1 yields

$$\frac{C_2}{C_1} = \frac{(1-k+kp)(1-kd)}{(1-0.8k)(1-k)}.$$

Harberger then assumed a 52-percent rate of taxation along with values of p of 0.275, 0.35, and 0.5 and values of d of 0.5, 0.65, and 0.75. He obtained nine estimates of the ratio C_2/C_1; these varied between 1.36 and 1.95, with some slight bias toward the lower figures. The average estimate was 1.62. These figures indicate that to obtain an equivalent income stream, between 1.36 and 1.95 times as many resources will be used in exploration and investment by mineral-producing companies as would be used in ordinary business investment (or as would be the case without depletion).

Harberger also showed that the ratio C_2/C_1 increases greatly, assuming specific values of d and p, as the rate of taxation increases. For example with $d = 0.65$ and $p = 0.35$, C_2/C_1 is 1.056 when $k = 0.1$, whereas C_2/C_1 is 1.519 when $k = 0.5$. We might note in passing that the firm engaged in mineral exploration is assumed to have other returns against which 80 percent of costs can be used as offsets. One must infer from Harberger's estimates that investment in oil exploration would be between 36 and 95 percent greater with percentage depletion and cost offsets than would be the case under ordinary depreciation and no offsets.

AN EMPIRICAL EXAMINATION OF THE OIL INDUSTRY

We now build on Harberger's results to consider the impact of the percentage depletion allowance on output and price, using data drawn from the United States oil industry. For our analysis we employ the period 1951–71. This period was selected for several reasons. First, it was a relatively stable period in the sense that there was little fluctuation in the price of crude petroleum. Second, the rate of depletion remained at 27.5 percent throughout the majority of this period and was lowered slightly only at the very end. Finally, and most important, this period would more closely approximate a period of equilibrium in the oil industry; during most of these years, the United States was not significantly dependent upon foreign oil. Further, the use of this period avoids the influence of governmental controls,[6] which have clearly had a diequilibrating effect.

The empirical analysis will be divided into two parts. Since our model is based on the assumption that the industry is reasonably competitive, we will first present evidence concerning the competitiveness of the petroleum industry. This will follow closely the approach of Chapter 3 in which we considered the competitiveness of the mining industry. Then, we will turn to our primary concern and provide some estimates of the impact of the percentage depletion allowance on output and price in the petroleum industry.

Competitiveness of the Petroleum Industry

The reader will recall that the model is based upon the assumption that, regardless of differences in rates and types of taxation among industries, competition will tend to equalize rates of return on investment, neglecting, of course, differences in risks. But if some industries for some reason or another receive a more lenient tax treatment, after-tax equality of rates of return require the before-tax rates to be lower in the favored industries relative to the unfavored. That is, favourable taxation in an industry causes increased investment in that industry until after-tax returns are equalized in the long run, thereby driving down the before-tax returns.

We shall examine the petroleum industry to see how well it fits this description. There are two problems that make this analysis difficult. In the first place, we were unable to obtain precise estimates of before- and after-tax rates of return for the oil industry and manufacturing. Second, there is really no precise statistical estimate of profitability before and after taxation. Therefore, we shall examine two different estimates of profitability to test our hypothesis. In each of these we will compare the before- and after-tax rates of return for Standard Poor's 425 industrials and on oil industry composite.*

Our first estimate of comparative returns to capital investment is the ratio of before- and after-tax operating profits to book value,† which reflects the relative returns to investors. Column 1 through 4 of Table 11.1 show these ratios for the oil composite and 425 industrials for each of the relevant years. Column 5 shows the difference in the ratios of the industrials and the oil composite before income taxes. As indicated at the end of the column, the mean difference is about 4.6 percent. The ratio for industrials is significantly higher than that for oil at the 1-percent level of significance (t = 12.11). In Figure 11.2 the two ratios are graphed for each year and the difference is apparent. The bottom of column 6 in Table 11.1, which indicates the differences between the ratios of after-tax profits to book value, shows that the mean of the differences is only − 0.005, a mean not significantly different from zero at any relevant level (t = 0.996). Figure 11.3 shows the closeness of the relation between the after-tax yearly ratios for the two groups of firms.

Also, the percentage of before- and after-tax earnings to book value can be used as an indicator of returns to capital investment.‡ These percentages, before and after income taxation, for the two groups of firms, are set forth in Column 1 through 4 of Table 11.2. The mean of the differences before taxation, Column 5, is 2.36 percent. This mean difference is again statistically significant at the 1-percent level (t = 9.61), indicating that the before-tax average percentage for the industrial group is higher. The relative percentages are graphed in Figure 11.4, and difference is again apparent. Column 6 shows the after-tax differences in percentages; the mean of the differences, 0.111, does not differ significantly from zero at any relevant level of confidence (t = 0.546). The extreme closeness of these percentages is clearly shown in Figure 11.5.

Hence, for both of our measures of profitability, our hypothesis has been supported. Our results indicate that the after-tax rate of return in the petroleum industry is equal to that prevailing in the rest of the economy and would therefore indicate that the industry fits the competitive model. Further, the fact that the before-tax rate of return is consistently lower in the petroleum industry indicates that this competitive adjustment to equilibrium was accomplished by the depletion allowance drawing additional resources (i.e., investment) into the petroleum industry.

---

*The oil composite is composed of international and domestic integrated companies and crude producers. The net-of-taxation figures are net of income taxation only—that is, of course, the basic form of taxation in which we are interested.

†Operating profit consists of sales less cost of goods sold and administrative expense before depreciation is deducted. Book value is total of common stock, capital surplus, and retained earnings less treasury stock, intangibles, and the difference between the carrying value and liquidating value of preferred stock.

‡Earnings are defined as net income less preferred dividends and including savings due to common stock as equivalents.

TABLE 11.1

Ratio of Profit to Book Value for 425 Industrials and an Oil Composite, before and after Taxes

Year	425 Industrials (Yearly Return on Book Value)		Oil Composite (Yearly Return on Book Value)		Difference between Rates of Return	
	Before Taxes (1)	After Taxes (2)	Before Taxes (3)	After Taxes (4)	Before-Tax (1−3) (5)	After-Tax (2−4) (6)
1951	0.36	0.19	0.32	0.23	0.04	−0.04
1952	0.32	0.18	0.28	0.22	0.04	−0.04
1953	0.34	0.19	0.30	0.23	0.04	−0.04
1954	0.29	0.19	0.25	0.21	0.04	−0.02
1955	0.34	0.21	0.28	0.22	0.06	−0.01
1956	0.32	0.20	0.28	0.23	0.04	−0.03
1957	0.30	0.20	0.28	0.23	0.02	−0.03
1958	0.25	0.17	0.22	0.18	0.03	−0.04
1959	0.27	0.18	0.22	0.18	0.05	0.00
1960	0.26	0.17	0.22	0.18	0.04	−0.01
1961	0.25	0.17	0.21	0.17	0.04	0.00
1962	0.27	0.18	0.21	0.18	0.06	0.00

1963	0.28	0.19	0.23	0.18	0.05	0.01
1964	0.29	0.20	0.22	0.18	0.07	0.02
1965	0.30	0.21	0.23	0.19	0.07	0.02
1966	0.32	0.22	0.25	0.20	0.07	0.02
1967	0.30	0.21	0.25	0.20	0.05	0.01
1968	0.32	0.22	0.26	0.20	0.06	0.02
1969	0.32	0.22	0.27	0.20	0.05	0.02
1970	0.30	0.21	0.27	0.19	0.03	0.02
1971	0.31	0.22	0.30	0.21	0.01	0.01
Mean	0.300	0.197	0.255	0.200	0.0457	−0.005
Standard deviation	0.030	0.017	0.033	0.020	0.0173	0.023

Source: Standard and Poor's *Analyst's Handbook*.

FIGURE 11.2

Ratio of Profits to Book Value before Taxes

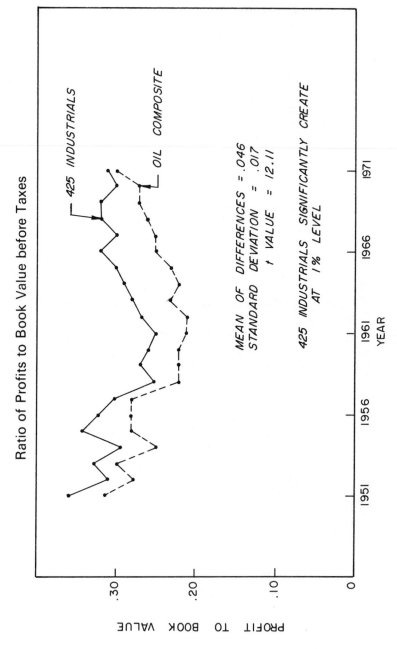

FIGURE 11.3

Ratio of Profits to Book Value after Taxes

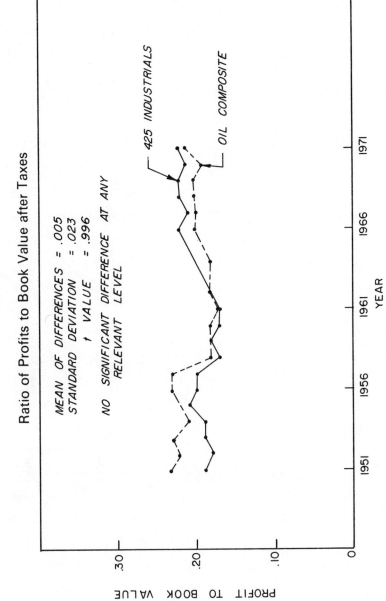

TABLE 11.2

Percentage of Earning to Book Value for 425 Industrials and an Oil Composite, Before and After Income Taxes

Year	425 Industrials (Yearly Return on Book Value)		Oil Composite (Yearly Return on Book Value)		Difference between Rates of Return	
	Before Taxes (1)	After Taxes (2)	Before Taxes (3)	After Taxes (4)	Before-Tax (1 − 3) (5)	After-Tax (2 − 4) (6)
1951	31.30	13.50	23.86	15.11	7.45	−1.61
1952	25.64	12.16	19.19	13.47	6.45	−1.31
1953	26.59	12.38	20.05	13.85	6.54	−1.47
1954	22.54	12.18	17.20	12.56	5.34	−0.38
1955	27.14	14.27	18.80	13.23	8.34	1.04
1956	24.52	13.28	18.76	13.58	5.76	−0.30
1957	21.74	11.99	17.48	12.80	4.26	−0.81
1958	17.45	9.62	12.83	9.20	4.62	0.42
1959	20.02	10.76	13.41	9.36	6.61	1.40
1960	18.58	10.08	13.58	9.67	5.00	0.41
1961	17.30	9.67	13.38	9.76	3.92	−0.09
1962	19.22	10.53	13.86	10.06	5.36	0.47

Year						
1963	20.30	11.11	15.07	10.80	5.23	0.31
1964	21.25	12.06	13.03	10.72	8.22	1.34
1965	22.16	12.64	15.89	11.07	6.27	1.57
1966	22.42	12.88	17.01	11.54	5.41	1.34
1967	20.36	11.76	17.59	11.64	2.77	0.12
1968	22.51	12.27	18.07	11.84	4.44	0.43
1969	21.80	11.86	18.39	11.72	3.41	0.14
1970	18.31	10.28	18.46	10.78	-0.15	-0.50
1971	19.81	10.80	21.01	10.99	-1.20	-0.19
Mean	21.95	11.71	17.00	11.61	4.95	0.111
Standard deviation	3.50	1.28	2.95	1.63	2.36	0.932

Source: Standard and Poor's *Analyst's Handbook.*

FIGURE 11.4

Ratio of Earnings to Book Value before Taxes

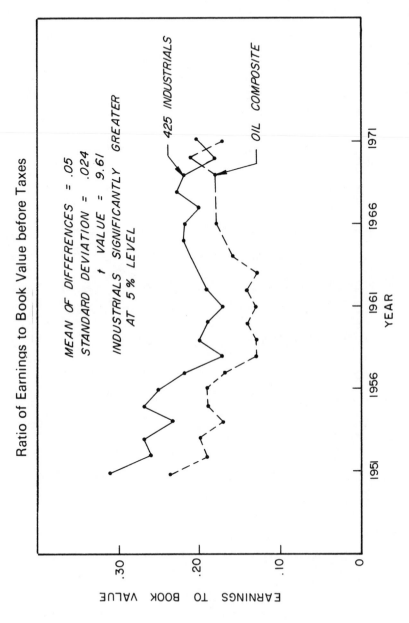

FIGURE 11.5

Ratio of Earnings to Book Value after Taxes

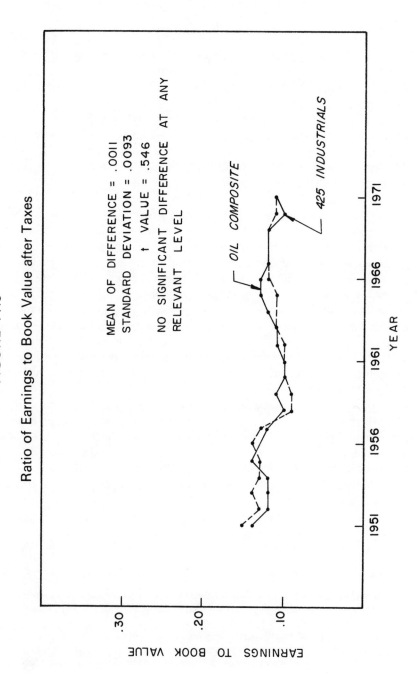

FIGURE 11.5

The Impact of the Percentage Depletion Allowance

At the outset, let us reemphasize the methodology to be employed in our estimation of the impact of the percentage depletion allowance on the oil industry.* Using a procedure akin to that employed previously in this book, we will perform a conceptual experiment. We observe what actually occurred over the 21-year period under the existing tax structure that included percentage depletion in segments of the oil industry. We then use our model and simulate the values of the important variables over the same period assuming that the oil industry had been subject to the same formula for tax depreciation as was manufacturing as a whole. The impact of percentage depletion would then be reflected in the difference between the actual and simulated values.

The model developed in the preceding section of this chapter predicted that the same discounted stream of income would elicit between 136 and 195 percent of the investment in an industry subject to the percentage depletion package relative to an industry using normal depreciation. Let us assume that these two figures bound the limits of response. These two limits are derived under the assumption of a 52-percent corporate income tax with a 0.275 depreciation allowance on *net—not gross—*income (the minimum) for the 136-percent estimates and a net depreciation allowance of 0.5 (the legal maximum) for the higher estimate. The assumed *total* discount rate is 0.75 for the 136, and 0.5 for the 195 estimate. The average of the nine estimates is approximately 62 percent more investment in oil relative to alternative investments with equal streams of returns.†

For our estimates of the price and output effects of depletion we shall use these three figures for our high, low, and middle estimates of the effect of percentage depletion on investment. We shall assume, for the reasons set forth above, that percentage depletion and the associated tax package were repealed at the beginning of 1951. We then simulate the effect of this repeal over the next 21 years, using the assumption that the oil industry operates under ordinary depreciation.

---

*By "depletion allowances" we imply the entire tax package as it applies specifically to oil. In such a package we include, as do others who have done research in this area, such items as expensing of intangibles.

†The reader will recall that the depletion figure used here is for income net of costs. The gross allowable is 0.275 and is therefore the minimum net allowable; the maximum net proportion allowed is 0.5. The discount factor is not the rate of interest but 9 total discount factor reflecting the difference between the discounted and nondiscounted stream of returns.

Since the above theoretical model emphasized that exploration and development investments are increased by depletion, we make an additional assumption. We assume that if investment, particularly exploration and development investment, is increased, the number of new producing wells increases at approximately the same rate. That is, if, as is assumed, a depletion allowance increases investment by 36 percent per year, the new wells that actually produce—not dry holes—increase by 36 percent also. This additional assumption appears to be justified empirically. Table 11.3 shows for each year from 1951 to 1971 the real exploration expenditures and the number of new

TABLE 11.3

Relation of Real Exploration Expenditures to New Producing Wells

Year	Exploration Expenditures (million dollars) Discounted by Finding Cost Index (1)	New Producing Wells (2)
1951	1,650	23,437
1952	1,620	25,448
1953	2,190	25,748
1954	2,346	29,776
1955	1,627	31,540
1956	2,190	31,196
1957	2,584	28,012
1958	1,553	24,578
1959	1,089	25,800
1960	1,331	21,186
1961	981	21,101
1962	1,507	21,249
1963	1,217	20,288
1964	1,326	20,620
1965	1,615	18,761
1966	963	16,780
1967	1,282	15,326
1968	*	*
1969	1,117	13,284
1970	1,238	12,230
1971	1,040	11,207

*1968 omitted because of north slope expenditures.

Source: Fred M. Howell and H. A. Merklein, "What It Costs to Find Hydrocarbons in the U.S.," World Oil (October 1973): 75–79.

producing wells. By *real exploration expenditures* we mean the total dollar amount of expenditures discounted by a finding cost index. The correlation between these is 0.709, a figure that is significant at the 1-percent level. Further, a simple logarithmic regression indicated that these data are consistent with our assumption of equal proportionate changes.*

We also make two additional assumptions about producing new wells. First, we assume that a new producing well has a life of approximately 20 years. Second, we assume for simplicity that new producing wells average about the same rate of production as the average of all producing wells. These assumptions seem somewhat reasonable and certainly aid in analytical simplicity.

Our method of analysis will be to estimate the number of producing new wells that would have been started in the absence of depletion. Suppose that 36 percent is the correct increase for the rate of investment—and by assumption for the increase in producing new wells—with depletion relative to regular depreciation. Consider column 2 of Table 11.4, which shows producing new wells in each year. In 1951, 23,437 new wells began producing. If N is the number of wells that would have begun producing in the absence of depletion, 1.36N = 23,437 or N = 23,437/1.36. Subtracting N from 23,437 we obtain 6,210, which is the reduction in the number of wells without depletion.† Column 3 presents the reduction in producing new wells under the 136-percent estimate for the effect of depletion. For each year we subtracted this reduction from the change in total wells (column 4) to obtain an estimate of total wells without depletion (shown in column 5). Column 6 shows the ratio of the two totals.

Using the assumptions that new wells average the same output as existing wells, we apply this percentage figure to actual output (column 1) in order to estimate the total output of crude petroleum without depletion (column 7). Column 8 shows the estimated reduction in output based upon the 1.36 estimate of the depletion effect. Note in column 8 the way in which this simulated reduction increases over time. At first the effect is quite small, but after ten years this reduction exceeds a 15-percent deficit.

---

*Using the data in Table 11.3, we estimated the equation $\ln NW = \alpha + \beta \ln E$, where NW is new producing wells and E is discounted exploration expenditures. In this estimating equation our assumption of equal proportionate changes requires that β be equal to one. We estimated β to be 0.70711 with a standard error of 0.17791. Hence, it is impossible to reject our hypothesis that β is equal to one at even the 90-percent confidence level. However, it should be noted that these results do not necessarily imply causality.

†Any differences are due to rounding. In our calculations we actually used R = PNW(0.36/1.36), where R is the reduction and PNW is the figure used in column 3.

TABLE 11.4

Output and Price Comparisons Using 1.36 Depletion Figure

Year	Output (1,000 bbls.) (1)	Producing New Wells (2)	Reduction in New Wells without Depletion (3)	Total Wells (4)
1951	2,247,711	23,437	6,210	474,990
1952	2,289,836	25,448	6,744	488,520
1953	2,357,082	25,748	6,823	498,940
1954	2,314,988	29,776	7,891	511,200
1955	2,484,428	31,540	8,363	524,010
1956	2,617,283	31,196	8,267	551,170
1957	2,616,9〔1	28,012	7,423	569,273
1958	2,449,061	24,578	6,513	574,905
1959	2,574,590	25,800	6,837	583,141
1960	2,574,933	21,186	5,614	591,158
1961	2,261,758	21,101	5,592	594,917
1962	2,676,189	21,249	5,631	596,385
1963	2,752,723	20,288	5,376	588,657
1964	2,786,822	20,620	5,464	588,225
1965	2,848,514	18,761	4,972	589,203
1966	3,027,763	16,780	4,441	583,302
1967	3,215,742	15,326	4,061	573,159
1968	3,290,042	14,227	3,770	553,920
1969	3,371,751	13,284	3,520	542,227
1970	3,319,945	12,230	3,241	530,990
1971	3,256,110	11,207	2,970	517,318

Source: Petroleum Facts and Figures, 1971 Edition (Washington, D.C.: American Petroleum Institute, 1971)

Figure 11.6 shows the effect more vividly. The upper line is the actual output per year, and the lower line is the simulated output without depletion. The difference in the simulated deficit is the distance between the two curves. Note the significant increase in the difference over time. This deficit is only 8 percent (209,883,000 bbls.) after five years. Thereafter, the deficit increases each five years to 13 percent (294,028,000) and 19 percent (575,275,000) until 1971, when the deficit is 25 percent, almost 1 billion barrels.

As output falls over time, price would have to increase, the rate of increase depending upon the elasticity of demand for crude petroleum. Most estimates

TABLE 11.4 *(Continued)*

Year	Total Wells without Depletion (5)	Percentage of Wells without Depletion (6)	Total Output without Depletion (1,000 bbls.) (7)	Reduction in Output (1,000 bbls.) (8)
1951	468,780	0.99	2,225,234	22,477
1952	475,566	0.97	2,221,140	68,695
1953	479,163	0.96	2,262,796	94,283
1954	483,532	0.95	2,199,939	115,749
1955	487,979	0.93	2,310,518	173,910
1956	506,872	0.92	2,407,900	209,383
1957	517,552	0.91	2,381,380	235,521
1958	516,671	0.90	2,204,155	244,906
1959	518,070	0.89	2,291,385	283,205
1960	520,473	0.88	2,265,941	309,992
1961	518,640	0.87	1,967,729	294,028
1962	504,477	0.85	2,274,761	401,428
1963	491,373	0.83	2,284,760	467,963
1964	485,477	0.83	2,313,062	473,759
1965	481,483	0.82	2,335,781	512,733
1966	471,142	0.81	2,452,488	575,275
1967	456,938	0.80	2,572,594	643,148
1968	433,929	0.78	2,566,233	723,809
1969	418,716	0.77	2,596,248	775,503
1970	403,238	0.76	2,523,158	796,787
1971	336,596	0.75	2,442,083	814,028

of elasticity vary between one (meaning a 1-percent decrease in quantity for a 1-percent increase in price) to 0.2 (meaning a 1-percent decrease in quantity for a 5-percent increase in price). We use these two bounds along with an intermediate estimate of 0.5 (a 1-percent decrease in quantity and a 2-percent increase in price). In Table 11.4 column 9 shows the actual price at the well, while columns 10 through 12 show the simulated price without depletion, using each of the three estimates of elasticity. Clearly, the lower the demand elasticity, the higher the simulated price in each year. As is the case for output, price increases very little initially. After about seven years the price increase becomes rather substantial.

Figure 11.7 shows the effects graphically. In ten years, the price of petroleum in the absence of depletion would have risen to between 15 and 75

		Price without Depletion		
Year	Price at Well (9)	Elasticity = 1 (10)	Elasticity = 0.5 (11)	Elasticity = 0.2 (12)
1951	$2.53	$2.56	$2.58	$2.65
1952	2.53	2.61	2.68	2.91
1953	2.68	2.79	2.89	3.22
1954	2.77	2.91	3.05	3.46
1955	2.77	2.96	3.16	3.74
1956	2.77	2.99	3.21	3.88
1957	3.09	3.37	3.65	4.48
1958	3.01	3.31	3.61	4.52
1959	2.90	3.22	3.54	4.50
1960	2.88	3.23	3.57	4.61
1961	2.89	3.27	3.64	4.77
1962	2.90	3.34	3.77	5.08
1963	2.89	3.38	3.87	5.35
1964	2.89	3.38	3.87	5.35
1965	2.86	3.37	3.89	5.43
1966	2.88	3.43	3.97	5.62
1967	2.88	3.46	4.03	5.76
1968	2.94	3.59	4.23	6.17
1969	3.09	3.80	4.51	6.64
1970	3.18	3.94	4.71	7.00
1971	3.41	4.26	5.12	7.67

percent higher than the actual price, depending on the elasticity assumed. As Figure 11.7 shows, the increase would have been between 19 and 95 percent after 15 years.

Table 11.4 and Figures 11.6 and 11.7 show the effect of depletion removal upon price and quantity under the smallest estimate of the effect of depletion on increased investment, 36 percent. Tables 11.5 and 11.6 and Figures 11.8 through 11.11 show the same effects under the average estimate of the increased investment, 62 percent, and the highest estimate, 95 percent. Each column in Tables 11.5 and 11.6 indicates the same variables as the equivalent column in Table 11.4. Of course, the simulated values differ more from the actual values with the higher estimates of the investment effect.

FIGURE 11.6

Output Deficit per Year Using 1.36 Depletion Figure

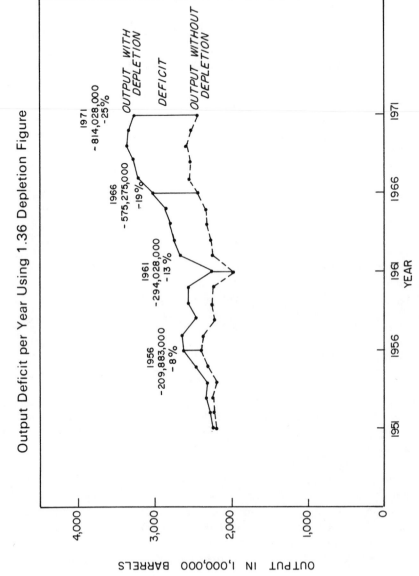

FIGURE 11.7

Price under Various Elasticities at 1.36 Estimate

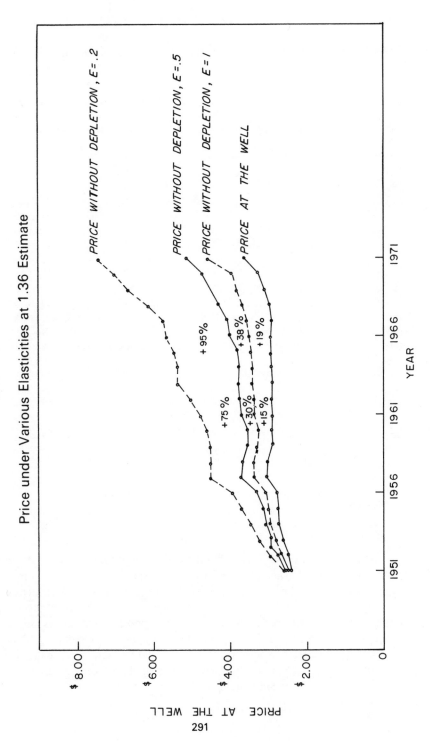

TABLE 11.5

Output and Price Comparisons Using 1.62 Depletion Figure

Year	Output (1,000 bbls.) (1)	Producing New Wells (2)	Reduction in New Wells without Depletion (3)	Total Wells (4)
1951	2,247,711	23,437	8,906	474,990
1952	2,289,836	25,448	9,670	488,520
1953	2,357,082	25,748	9,784	498,940
1954	2,314,988	29,776	11,315	511,200
1955	2,484,428	31,540	11,985	524,010
1956	2,617,283	31,196	11,854	551,170
1957	2,616,901	28,012	10,645	569,273
1958	2,449,061	24,578	9,340	574,905
1959	2,574,590	25,800	9,804	583,141
1960	2,574,933	21,186	8,051	591,158
1961	2,261,758	21,101	8,018	594,917
1962	2,676,189	21,249	8,075	596,385
1963	2,752,723	20,288	7,709	588,657
1964	2,786,822	20,620	7,836	588,225
1965	2,848,514	18,761	7,129	589,203
1966	3,027,763	16,780	6,376	583,302
1967	3,215,742	15,326	5,824	573,159
1968	3,290,042	14,227	5,406	553,920
1969	3,371,751	13,284	5,048	542,227
1970	3,319,945	12,230	4,647	530,990
1971	3,256,110	11,207	4,259	517,318

Source: Petroleum Facts and Figures, 1971. Edition (Washington, D.C.: American Petroleum Institute, 1971)

It can be seen in Figure 11.8 that the simulated output deficit is somewhat higher in each year as the estimate of the investment effect increases from 36 to 62 percent. In 1956 the output deficit was 12 percent; in 1961, 18 percent; and in 1966, 25 percent. By 1971 the effect was 33 percent, or a deficit of more than 1 billion barrels. The price effect with the 62-percent estimate is shown in Figure 11.9. It can be seen that again in the early stages the effect is small, but it rises rapidly thereafter to between an 18- and a 90-percent increase in price by 1961 and between a 25- and a 125-percent increase by 1966. It is apparent that the price effect is quite significant in the later years.

Year	Total Wells without Depletion (5)	Percentage of Wells without Depletion (6)	Total Output without Depletion (1,000 bbls.) (7)	Reduction in Output (1,000 bbls.) (8)
1951	466,084	0.98	2,202,757	44,954
1952	469,924	0.96	2,198,243	91,593
1953	470,560	0.94	2,215,659	141,423
1954	471,505	0.92	2,129,789	185,199
1955	472,330	0.90	2,235,985	248,443
1956	487,636	0.88	2,303,209	314,074
1957	495,094	0.87	2,276,703	340,197
1958	491,386	0.85	2,081,702	367,359
1959	489,818	0.84	2,162,656	411,934
1960	489,767	0.83	2,137,194	437,739
1961	485,508	0.82	1,854,642	407,116
1962	478,901	0.80	2,140,951	535,237
1963	463,464	0.79	2,174,651	578,072
1964	455,196	0.77	2,145,853	640,969
1965	449,045	0.76	2,164,871	683,643
1966	436,768	0.75	2,270,822	756,941
1967	420,801	0.73	2,347,492	868,250
1968	396,156	0.72	2,368,830	921,212
1969	379,415	0.70	2,360,225	1,011,525
1970	363,531	0.68	2,257,562	1,062,382
1971	345,600	0.67	2,181,594	1,074,516

Finally, Figures 11.10 and 11.11 indicate huge quantity deficits and price increases after about five years under the 95-percent increased investment estimate. The different estimates are clearly shown on the graph.

Our estimates indicate that the percentage depletion allowance had a substantial affect on the petroleum industry, even if one assumes the smallest impact on investment and the most elastic demand function. In the text we have merely attempted to summarize the results. A more precise analysis of the effects can be obtained by closer examination of the relevant tables and figures.

TABLE 11.5 *(Continued)*

Year	Price at Well (9)	Price without Depletion		
		Elasticity = 1 (10)	Elasticity = 0.5 (11)	Elasticity = 0.2 (12)
1951	$2.53	$2.58	$2.63	$2.78
1952	2.53	2.63	2.73	3.04
1953	2.68	2.84	3.00	3.48
1954	2.77	2.99	3.21	3.88
1955	2.77	3.05	3.32	4.16
1956	2.77	3.10	3.43	4.43
1957	3.09	3.49	3.89	5.10
1958	3.01	3.46	3.91	5.27
1959	2.90	3.36	3.83	5.22
1960	2.88	3.37	3.86	5.33
1961	2.89	3.41	3.93	5.49
1962	2.90	3.48	4.06	5.80
1963	2.89	3.50	4.10	5.92
1964	2.89	3.55	4.22	6.21
1965	2.86	3.55	4.23	6.29
1966	2.88	3.60	4.32	6.48
1967	2.88	3.66	4.44	6.77
1968	2.94	3.76	4.59	7.06
1969	3.09	4.02	4.94	7.73
1970	3.18	4.20	5.22	8.27
1971	3.41	4.54	5.66	9.04

Some Concluding Remarks

We should at this point clarify and explain some of the implicit assumptions upon which our empirical conclusions are based, because changing these assumptions could alter the estimates either upward or downward, depending upon which assumptions are changed. In the first place we have implicitly assumed that all producing new wells produce for approximately the average length of production. Clearly, some new wells do not last long. To the extent that some of the wells, simulated as not being drilled in the absence of depletion, did not in reality produce throughout the period, the estimates of the decreases in quantity and increases in price may be biased downward and upward, respectively, particularly toward the latter years of the period, when it is more likely that some of the wells were closed down. Offsetting to a greater or lesser degree this possible bias, however, is the

TABLE 11.6

Output and Price Comparisons Using 1.95 Depletion Figure

Year	Output (1,000 bbls.) (1)	Producing New Wells (2)	Reduction in New Wells without Depletion (3)	Total Wells (4)
1951	2,247,711	23,437	11,414	474,990
1952	2,289,836	25,448	12,393	488,520
1953	2,357,082	25,748	12,539	498,940
1954	2,314,988	29,776	14,501	511,200
1955	2,484,428	31,540	15,360	524,010
1956	2,617,283	31,196	15,192	551,170
1957	2,616,901	28,012	13,646	569,273
1958	2,449,061	24,578	11,969	574,905
1959	2,574,590	25,800	12,565	583,141
1960	2,574,933	21,186	10,318	591,158
1961	2,261,758	21,101	10,276	594,917
1962	2,676,189	21,249	10,348	596,385
1963	2,752,723	20,288	9,880	588,657
1964	2,786,822	20,620	10,042	588,225
1965	2,848,514	18,761	9,137	589,203
1966	3,027,763	16,780	8,172	583,302
1967	3,215,742	15,326	7,464	573,159
1968	3,290,042	14,227	6,929	553,920
1969	3,371,751	13,284	6,469	542,227
1970	3,319,945	12,230	5,956	530,990
1971	3,256,110	11,207	5,488	517,318

Source: *Petroleum Facts and Figures, 1971, Edition* (Washington, D.C.: American Petroleum Institute, 1971)

bias that could follow from our implicit assumption that no additional producing wells would have been shut down because the theoretical elimination of the depletion allowance made them unprofitable. This assumption would bias the quantity simulations upward and the price simulation downward. Thus, the consequences of these two assumptions are probably offsetting in large measure.

Second, we have assumed, as did Harberger in his original model, that in any one year the increased price of petroleum would not have elicited increased supply. That is, we have assumed the industry to be a constant-cost industry—an assumption that may not be particularly unrealistic over the time period considered.

TABLE 11.6 *(Continued)*

Year	Total Wells without Depletion (5)	Percentage of Wells without Depletion (6)	Total Output without Depletion (1,000 bbls.) (7)	Reduction in Output (1,000 bbls.) (8)
1951	463,576	0.98	2,220,757	44,954
1952	464,713	0.95	2,175,344	114,492
1953	462,594	0.93	2,192,086	164,996
1954	460,353	0.90	2,083,489	231,499
1955	457,803	0.87	2,161,452	322,976
1956	469,771	0.85	2,224,691	392,592
1957	474,228	0.83	2,172,028	444,873
1958	467,891	0.81	1,983,739	465,322
1959	463,562	0.79	2,033,926	540,644
1960	461,261	0.78	2,008,448	566,485
1961	454,744	0.76	1,718,936	542,822
1962	445,864	0.75	2,007,141	669,047
1963	428,258	0.73	2,009,488	743,235
1964	417,784	0.71	1,978,644	808,178
1965	409,625	0.70	1,993,960	854,554
1966	395,552	0.68	2,058,879	968,884
1967	377,945	0.66	2,122,390	1,093,352
1968	351,777	0.64	2,105,627	1,184,415
1969	333,615	0.62	2,090,486	1,281,265
1970	316,442	0.60	1,991,967	1,327,978
1971	297,282	0.57	1,855,983	1,400,127

Admittedly, in the short run the higher price resulting from the hypothetical elimination of the depletion allowance would have induced entry, but this entry would have driven price downward below the long-run average cost, including the going return on capital. The losses thereby sustained would have driven some firms out of business and caused quantity to be reduced and price to be driven back up. The relatively constant petroleum price along with the increased production during a period of mild inflation provides some evidence that crude petroleum production behaved somewhat as a constant cost industry over this period. However, to the extent that the long-run supply had something less than infinite elasticity, our estimates of the decreases in quantity are biased downward, and those of the increases in price are biased upward.

Next, we must mention the effect of imports. Obviously our conclusions about output and price are based upon the assumption that imports would have remained unchanged during the period after the percentage depletion

Year	Price at Well (9)	Price without Depletion		
		Elasticity = 1 (10)	Elasticity = 0.5 (11)	Elasticity = 0.2 (12)
1951	$2.53	$2.58	$2.63	$2.78
1952	2.53	2.66	2.78	3.16
1953	2.68	2.87	3.06	3.62
1954	2.77	3.05	3.32	4.16
1955	2.77	3.13	3.49	4.57
1956	2.77	3.19	3.60	4.85
1957	3.09	3.62	4.14	5.72
1958	3.01	3.58	4.15	5.87
1959	2.90	3.51	4.12	5.95
1960	2.88	3.51	4.15	6.05
1961	2.89	3.58	4.28	6.36
1962	2.90	3.63	4.35	6.53
1963	2.89	3.67	4.45	6.79
1964	2.89	3.73	4.57	7.08
1965	2.86	3.72	4.58	7.15
1966	2.88	3.80	4.72	7.49
1967	2.88	3.86	4.84	7.78
1968	2.94	4.00	5.06	8.23
1969	3.09	4.26	5.44	9.25
1970	3.18	4.45	5.72	9.54
1971	3.41	4.49	6.34	10.74

allowance was removed. Certainly, the United States might well have chosen to counteract the decreased oil supply and consequent price increase by becoming increasingly dependent upon foreign oil supplies. This is, however, not the point at issue here. We wished to examine the effects of removal *per se* in isolation from those costs that society perhaps would have paid in order to ameliorate these effects. For this reason we chose not to allow imports to increase during the period of simulation. In any case, it would have been extremely difficult to predict what would have happened, particularly with import quotas and so forth, since we omitted political, as opposed to economic, decision variables from the model. Finally, of course, we have estimated the reduction in U.S. Production, which presumably would not have increased if imports had increased. In fact, to the extent that increased foreign oil had driven down oil prices there might possibly have been a further decrease in U.S. production.

FIGURE 11.8

Output Deficit per Year Using 1.62 Estimate

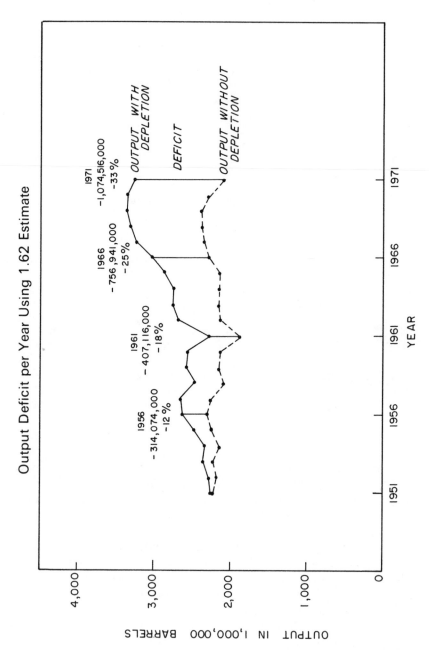

298

FIGURE 11.9

Prices under Various Elasticities at 1.62 Estimate

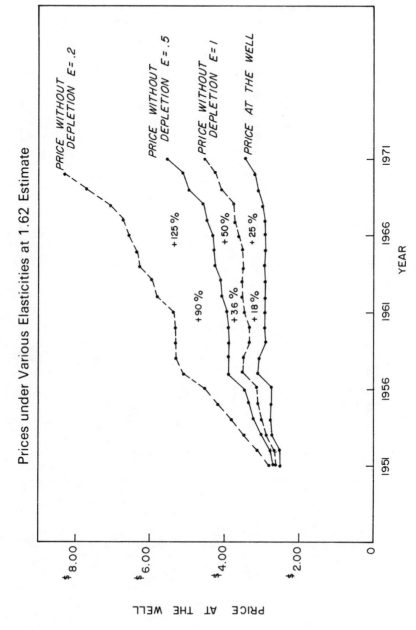

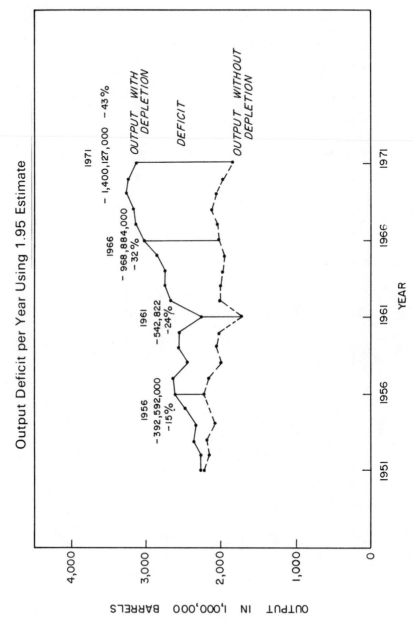

FIGURE 11.10

Output Deficit per Year Using 1.95 Estimate

300

FIGURE 11.11

Price under Various Elasticities at 1.95 Estimate

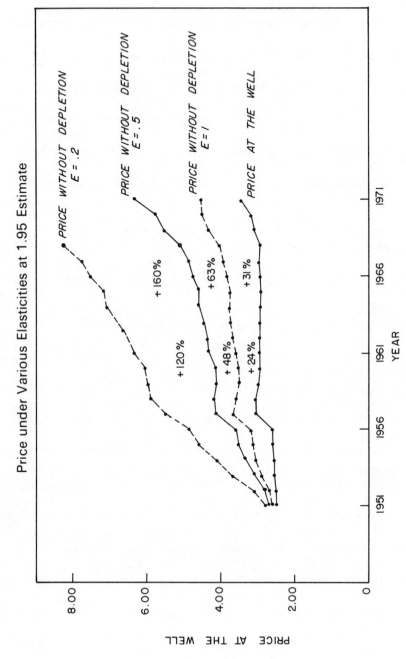

We might say a word about the effect upon reserves. To some extent the decreased output of oil that would have resulted from elimination of depletion would have increased present reserves. On the other hand, the resulting lower exploration and development expenditure would have decreased findings, and this diminution would have lowered present reserves. It is difficult to estimate the overall effect of these perhaps offsetting forces. This is more of a question for petroleum experts than for economists.

Given these possible biases and the simulation measures already presented, what can we conclude? This is a difficult question. First, let us address ourselves to the problem of deciding which estimate of the increased investment figure to use. In the first place, the estimate of a 36-percent increase in investment seems far too low, other things being equal, because the net depletion figure used to derive it seems much too low and the associated discount factor much too high. Any estimate above a 65- to 70-percent increase is probably too high. Without biases, the 62-percent figure would probably be rather accurate, but the biases discussed would probably have driven the quantity slightly downward. Thus, we would speculate a figure between the 36- and 63-percent estimate, but much closer to the higher figure. Our best estimate of the average demand elasticity during this period is a figure slightly above our 0.5 estimate. Thus, our price simulation is probably rather close to that obtained with the 62-percent investment increase and a 0.5 demand elasticity. Certainly, others might speculate different estimates, but our high and low simulation probably bound what would actually have occurred.

In any case every estimate shows a rather similar pattern. At first there was not a particularly significant effect after the simulated elimination (recall that we did not permit any oil wells to shut down). Then after about five years both the simulated quantity and the simulated price began to deviate widely from the price that actually existed. It is fairly clear that the process of adjustment is dynamic: there is not a process of immediate reaction to the change followed by attainment of a new long-run equilibrium.

Finally, we should again stress the fact that this analysis was designed to predict what *would* have happened in the period considered if the percentage depletion allowance had been eliminated. We therefore had to take other existing economic and political conditions as *given*. It follows then that any such approach might not be able to predict precisely what actually happened and is happening as a result of the elimination of the percentage-depletion allowance because the economic and, particularly, the politicial environment changed radically following 1971. This caveat notwithstanding, our results are important, since they indicate the *independent* effect of this type of tax treatment.

NOTES

1. Arnold C. Harberger, "The Taxation of Mineral Industries," *Federal Tax Policy for Economic Growth and Stability* (Washington D.C.: Government Printing Office, 1955). 439–49. The basic points in that original paper were developed further in Arnold C. Harberger, "The Tax Treatment of Oil Exploration," *Proceedings of the Second Energy Institute* (Washington, D.C.: The American University, 1961), 256–69. These papers are reprinted as Chapters 11 and 12 in Arnold C. Harberger, *Taxation and Welfare* (Boston: Little, Brown, 1974).

2. Stephen L. McDonald, "Percentage Depletion and the Allocation of Resources: The Case of Oil and Gas," *National Tax Journal* (December 1961): 323–36.

3. The following papers are some of the major ones in the controversy. Richard A. Musgrave, "Another Look at Depletion," *National Tax Journal* (June 1962): 205–8; Douglas H. Eldridge, "Rate of Return, Resource Allocation and Percentage Depletion," *National Tax Journal* (June 1962): 209–17; Stephen L. McDonald, "Percentage Depletion and Tax Neutrality: A Reply to Messrs. Musgrave and Eldridge," *National Tax Journal* (September 1962): 314–26; Peter O. Steiner, "The Non-Neutrality of Corporate Income Taxation—With and Without Depletion," *National Tax Journal* (September 1963): 238–51: McDonald's reply and Steiner's rejoiner were in the March 1964 *National Tax Journal.*

4. Paul Davidson, "Public Policy Problems of the Domestic Crude Oil Industry," *American Economic Review* (March 1963): 85–108.

5. For example, Ducan R.G. Campbell, "Public Policy Problems of the Domestic Crude Oil Industry: Comment," *American Economic Review* (March 1964): 114–18.

6. For a discussion of the industry and of governmental regulations, the cf. E. Anthony Copp, *Regulating Competition in Oil, Government Intervention in the U.S. Refining Industry, 1948–1975* (College Station: Texas A&M University Press, 1976).

Bibliography

Alchian, A. and H. Demsetz. "The Property Rights Paradigm". *Journal of Economic History* (March 1973): 16–27.

Allen, R.G.D. *Mathematical Analysis for Economists*. London, Macmillan, 1938.

Anders, G., W.P. Gramm, and S.C. Maurice. *The Impact of Taxation and Environmental Controls on the Ontario Mining Industry. – A Pilot Study* Toronto: Ontario Ministry of Natural Resources, 1975.

―――. *Does Resource Conservation Pay?* Los Angeles: International Institute for Economic Research, 1978.

―――, and C.W. Smithson. *Investment Effects on the Mineral Industry of Tax and Environmental Policy Changes: A Simulation Model*, Toronto: Ontario Ministry of Natural Resources, July 1978.

Arad, R.W., and U.B. Arad. "Scarce Natural Resources and Potential Conflict." In *Sharing Global Resources*. New York: McGraw-Hill, 1979.

Arrow, K.J. *Social Choice and Individual Values*. New York: Wiley, 1951.

Barnett, H.J., and C. Morse. *Scarcity and Growth*. The Johns Hopkins University Press, Baltimore, 1962.

Bennett, G.L. "The Impact of Environmental Controls and Taxes on Mining." Unpublished dissertation, Texas A&M University, May 1977.

Berndt, E.R., and L.R. Christensen. "The Internal Structure of Functional Relationships: Separability, Substitution, and Aggregation." *Review of Economic Studies* (July 1973): 403–10.

Berndt, E.R., and D.W. Jorgenson. "Production Structure." In D.W. Jorgenson, E.R. Berndt, L.R. Christensen, and E.A. Hudson. *U.S. Energy Resources and Economic Growth*, Final Report to the Ford Foundation Energy Policy Project, Washington, D.C., October 1973.

Berndt, E.R., and D.O. Wood. "Technology, Prices, and the Derived Demand for Energy." *Review of Economics and Statistics* (August 1975): 259–68.

―――. "Engineering and Econometric Approaches to Industrial Energy Conservation and Capital Formation: A Reconciliation." Unpublished working paper, University of British Columbia, December 1977.

Blackorby, C., D. Primont, and R.R. Russell. "On Testing Separability Restrictions with Flexible Functional Forms." *Journal of Econometrics* (March 1977): 195–209.

304

Buchanan, J.M., and G. Tullock. *The Calculus of Consent*. Ann Arbor: University of Michigan Press, 1965.

Campbell, D.R.G. "Public Policy Problems of the Domestic Crude Oil Industry: Comment." *American Economic Review* (March 1964): 114–18.

Carlisle, D. "The Economics of a Fund Resource with Particular Reference to Mining." *American Economic Review* (September 1954): 595–616.

Christensen, L.R., D. Jorgenson, and L. Lau. "Conjugate Duality and the Transcendental Logarithmic Production Function." Presented at the Second World Congress of the Econometric Society, Cambridge England, 1970.

Christensen, L.R., and W.H. Greene. "Economies of Scale in U.S. Electric Power Generation." *Journal of Political Economy* (August 1976): 665–76.

Coase, R. "The Problem of Social Cost." *Journal of Law and Economics* (October 1960): 1–44.

Copp, E.A. *Regulating Competition in Oil, Government Intervention in the U.S. Refining Industry, 1948–1975*. College Station: Texas A&M University Press, 1976.

Council on Environmental Quality. *The Economic Impact of Pollution Control: A Summary of Recent Studies*. Washington, D.C.: Government Printing Office, 1972.

Cummings, R.L. "Some Extensions of the Economic Theory of Exhaustible Resources." *Western Economic Review* (September 1969): 201–10.

Davidson, P. "Public Policy Problems of the Domestic Crude Oil Industry." *American Economic Review* (March 1963): 85–108.

Davis, O.A., and A.B. Whinston. "Economics of Urban Renewal." *Law and Contemporary Problems* (Winter 1961): 105–17.

Denny, M., and M. Fuss. "The Use of Approximation Analysis to Test for Separability and the Existence of Consistent Aggregates." *American Economic Review* (June 1977): 404–18.

———, J.D. May, and C. Pinto. "The Demand for Energy in Canadian Manufacturing: Prologue to an Energy Policy." *Canadian Journal of Economics* (May 1978): 300–13.

Eldridge, D.H. "Rate of Return, Resource Allocation and Percentage Depletion." *National Tax Journal* (September 1962): 209–17.

Ferguson, C.E. and S.C. Maurice. *Economic Analysis*. Homewood, Ill.: Irwin, 1978.

Frost, S. "Yes—But What Are the Facts?" *The Outlook* (March 24, 1924; April 2, 1924).

Furubotn, E.G. "The Orthodox Production Function and the Adaptability of Capital." *Western Economic Journal* (Summer 1965): 288–300.

———. "Long-Run Analysis and the Form of the Production Function." *Economia Internazionale* (February 1970): 3–35.

Furubotn, E.G., and S. Pejovich (eds.). *The Economics of Property Rights.* Cambridge, Ballinger, Mass. 1974.

Fuss, M.A. "The Demand for Energy in Canadian Manufacturing: An Example of the Estimation of Production Structures with Many Inputs." *Journal of Econometrics* (January 1977): 89–116.

Gaffney, M (ed.). *Extractive Resources and Taxation.* Madison: University of Wisconsin Press, 1967.

Gordon, R.L. "A Reinterpretation of the Pure Theory of Exhaustion." *The Journal of Political Economy* (June 1967): 274–86.

Gramm, W.P. "A Theoretical Note on the Capacity of the Market System to Abate Pollution." *Land Economics* (August 1969): 336–38.

——. "The Energy Crisis in Perspective." *Wall Street Journal* (Friday, November 30, 1973).

——. "Debunking Doomsday." *Reason* (August 1978).

——, and S.C. Maurice. "The *Real* Teapot Dome Scandal . . . Crude Stockpiling: A Prime Example of Poor Economics." *World Oil* (November 1975): 71–73.

——. "The Economic Feasibility of Peacetime Nuclear Explosives in Mineral Extraction." In *PNE (Peaceful Nuclear Explosives) Activity Projections for Arms Control Planning.* A report submitted to the U.S. Arms Control and Disarmament Agency by Gulf Universities Research Consortium, January 20, 1975.

Gray, L.C. "Rent under the Assumption of Exhaustibility." *Quarterly Journal of Economics* (May 1914): 464–89.

Griffin, J.M. and P.R. Gregory. "An Intercountry Translog Model of Energy Substitution Responses." *American Economic Review* (December 1976): 845–57.

Harberger, A.C. "The Taxation of Mineral Industries." *Federal Tax Policy for Economic Growth and Stability.* Washington, D.C.: Government Printing Office, 1955.

——. "The Tax Treatment of Oil Exploration." *Proceedings of the Second Energy Institute.* Washington, D.C.: The American University, 1961.

——. *Taxation and Welfare.* Boston: Little, Brown, 1974.

Haveman, R.V. "Common Property, Congestion, and Environmental Pollution." *Quarterly Journal of Economics* (May 1973): 278–87.

Herfindahl, O.C. "Some Fundamentals of Mineral Economics." *Land Economics* (May 1955): 131–38.

——. "Depletion and Economic Theory." In M. Gaffney (ed.), *Extractive Resources and Taxation.* Madison: University of Wisconsin Press, 1967, 63–90.

Hogan, J.D. "Resource Exploitation and Optimum Tax Policies: A Control Model Approach." In M. Gaffney (ed.), *Extractive Resources and Taxation*. Madison: University of Wisconsin Press, 1967, 91–106.

Hotelling, H. "The Economics of Exhaustible Resources." *The Journal of Political Economy* (April 1931): 137–75.

Howell, F.M., and Merklein, H.A. "What It Costs to Find Hydrocarbons in the U.S." *World Oil* (October 1973): 75–79.

Humphrey, D.B., and J.R. Moroney. "Substitution among Capital, Labor, and Natural Resource Products in American Manufacturing." *Journal of Political Economy* (February 1975): 57–82.

Jorgenson, D.W., and Z. Griliches. "The Explanation of Productivity Change." *The Review of Economic Studies* (July 1967): 249–84.

Kmenta, J., and R. Gilbert. "Small Sample Properties of Alternative Estimators of Seemingly Unrelated Regressions." *Journal of the American Statistical Association* (December 1968): 1180–1200.

Lahiri, K. "Inflationary Expectations: Their Formation and Interest Rate Effects." *American Economic Review* (March 1976): 124–31.

Marshall, A. *Principles of Economics*. London: Macmillan, 1920.

McCulloch, R., and J. Piñera, "Alternative Commodity Trade Regimes." In *Sharing Global Resources*. New York: McGraw-Hill, 1979.

McDonald, S.L. "Percentage Depletion and the Allocation of Resources: The Case of Oil and Gas." *National Tax Journal* (December 1961): 323–36.

———. "Percentage Depletion and Tax Neutrality: A Reply to Messrs. Musgrave and Eldridge." *National Tax Journal* (September 1962): 314–26.

Meadows, D.H., D.L. Meadows, J. Randers, and W.H. Behrens III. *The Limits to Growth*. New York: Universe, 1972.

Musgrave, R.A. "Another Look at Depletion." *National Tax Journal* (June 1962): 205–8.

Peterson, F.M., and A.C. Fisher. "The Exploitation of Extractive Resources: A Survey" *The Economic Journal* (December 1977): 681–721.

Pigou, A.C. *The Economics of Welfare*. London: Macmillan, 1929.

Samuelson, P.A. *Foundations of Economic Analysis*. Cambridge, Mass.: Harvard University Press, 1947.

Scott, A.T. "The Theory of the Mine under Conditions of Certainity." In M. Gaffney (ed.), *Extractive Resources and Taxation*. Madison: University of Wisconsin Press, 1967, 25–62.

Shephard, R.W. *Theory of Cost and Production Functions*. Princeton: Princeton University Press, 1970.

Smith, V.L. "Economics of Production from Natural Resources." *American Economic Review* (June 1968): 409–31.

Smithson, C.W., G. Anders, W.P. Gramm, and S.C. Maurice. *Factor Substitution and Biased Technical Change in the Canadian Mining Industry*. Ontario, Canada: Ministry of Natural Resources, 1979.

Solow, R.M. "The Economics of Resources or the Resources of Economics." *American Economic Review* (May 1974): 1–14.

———— and R. Wan. "Extration Costs in the Theory of Exhaustible Resources." *The Bell Journal of Economics* (Autumn 1976): 359–70.

Sweeney, J. "Economics of Depletable Resources: Market Forces and Intertemporal Bias." *Review of Economic Studies* (February 1977): 125–42.

Steel, H. "Natural Resource Taxation: Resource Allocation and Distributional Implications." In M. Gaffney (ed.), *Extractive Resources and Taxation*. Madison: University of Wisconsin Press, 1967, 233–68.

Steiner, P.O. "The Non-Neutrality of Corporate Income Taxation—With and Without Depletion." *National Tax Journal* (September 1963): 238–51.

Tower, W.S. *A History of the American Whale Fishery*. Philadelphia: Winston, 1907.

Uzawa, H. "Production Functions with Constant Elasticities of Substitution." *Review of Economic Studies* (October 1962): 291–99.

Weitzman, M. "The Optimal Development of Resource Pools." *Journal of Economic Theory* (June 1976): 351–64.

About the Authors

GERHARD ANDERS, Senior Policy Advisor, Ontario Ministry of Natural Resources, Supervisor of that Ministry's Metallic Minerals Section, and Adjunct Associate Professor, Department of Geological Engineering, University of Toronto, is a Mining Engineer and Economist now specializing in Mineral Economics and Policy Analysis. Before joining the ministry, Dr. Anders taught Economics in Florida, headed the Economic Survey program mounted by the Canadian federal government in Canada's Northwest Territories, and spent many years in the Canadian mining industry in various senior technical and managerial capacities. Dr. Anders received his Ph.D. in Economics from Texas A&M University and a Dipl. Ing. (M.Sc.) from the Technical University, Clausthal, West Germany. He has published and coauthored numerous articles and monographs on both engineering and economic subjects. Dr. Anders is also a member of the editorial board of the Fisher Institute, Dallas, Texas.

The Honorable W. PHILIP GRAMM, member of the U.S. House of Representatives and former professor of economics at Texas A&M University, received his Ph.D. from the University of Georgia in 1967. He has written more than 40 articles, books, and monographs, covering the spectrum of economics from environment and energy to banking and inflation. Dr. Gramm has served as a consultant on energy, environment, and taxation to the Ministry of Natural Resources of the Canadian government. He served as a member of the Gulf Universities Research Consortium, and has testified before both houses of Congress and several state legislatures on energy problems.

Dr. S. CHARLES MAURICE is professor of economics and head of the department of economics at Texas A&M University. He joined the University in 1967. He has published extensively in microeconomic theory, mineral economics, and industrial organization and has a best-selling textbook in microeconomic theory. He has been consultant to the U.S. government and to Ontario Ministry of Natural Resources in matters concerning energy forecasting and resource conservation. He has testified before United States Senate Committee's concerning these matters.

CHARLES W. SMITHSON is Assistant Professor of Economics at Texas A&M University. After receiving his Ph.D. from Tulane University, he joined the faculty in 1976. Dr. Smithson has been actively engaged in research in the areas of regulation, production, and natural resources—particularly with the question of input substitutability—and has written several articles and monographs in these areas.

Index

Arab oil embargo, 229–30
Assets
    markets in equilibrium, 48; to analyze
    multi-grade deposits, 48–49; value in
    relation to discount rate, 231–32

Benefits,
    marginal, social, 128–29; of environ-
    mental controls, 202–27; effect on
    human capital, 208–10; effect on
    mineral firm employees, 217–26;
    effect on property owners, 204–07;
    estimation, 210–17; measured in agri-
    cultural land use, 210–14; measured
    in forestry land, 213–16; theory of,
    202–05; in relation to mining, 204–10
Berndt, Ernst R., 65–78
Bonus payment 30–32; See also Royalty
Book value, 57–60

Capital
    cost, 100–01; equipment, 69–70; flow
    of, 58–59; goods, 146–47; human,
    208–10; investment, 147–49; owners
    of, 204–07; productivity, 246; ratio of
    profits to, 52–57, 190–91, 194,
    196–97; return, 114; services, prices
    of, 67–69; stock, 146–47
Carlisle, Donald, 38–41
Christensen, Laurits R., 66–78
Common property resource, 47–48
Competition, 30–32, 37–38, 94–97, 122,
    137–38; and excess returns, 155; and
    investment, 158–59, 165–66, 266–
    67; and net worth, 151–52; empirical
    evidence, 50–62; equilibrium, 273; in
    petroleum industry, 267–75, 284;
    setting market rate, 246; 272–73
Complementarity of inputs, 65–66, 72–73
Conservation, 228–63; and interest rates,
    253–60; as government policy,
    238–42; cost of, 238–40; in relation to
    investment value, 242–48; long-term,
    240–41; movement in U.S., 229–30;
    social costs, 253–60; Teapot Dome
    affair, 253–60; versus yield on in-
    vestment, 242–49

Constraints, externally imposed, See
    Environmental controls
Consumer price index, 246, 244–48,
    248–49, 252
Corporate bonds, 246–48, 248, 250–52,
    259
Cost
    affected by environmental controls,
    taxation, 141–69; capital, See Cost,
    fixed; change in, 111–12, 114–16,
    117–19, 120–21; discounted future
    difference between price and cost,
    232, 234; effect of output on, 16, 18–
    19; externality, 128–29; fixed, 91–94,
    100–01, 113–14; future, 111; hazard,
    100–01, 113–14; in relation to return
    rate, 51; increasing, 117, 119, 128,
    142–43; land, 100; marginal, 41–42,
    43–44, 102–03, 104–05, 108–09; af-
    fecting output, 143–44; definition of,
    13–14; marginal reduction in,
    110–11; of environmental controls,
    128, 134–35, 137–39, 160–61, 177–
    79, 181, 200; of exploration, 40–41; of
    extraction, 19–20, 32–33, 91–94; of
    information, 137; of investment,
    151–52, 154–56, 157–58, 160–61; of
    new technology, 89–90, 104–05; of
    production, 27–28, 234–37; oper-
    ating, See Cost, variable; social, See
    Externalities; total, 94–97; variable,
    91–94, 100, 102–03, 108–09,
    110–11, 113–14, 154–55
Cost functions, 81–83; analysis of, 13–15;
    change in, 102–03; effect of level of
    extraction on, 101–02; in parametric
    form, 70–73; marginal, 104–05;
    marginal private, 128–29; marginal
    social, 128–29; of extraction, 14–15;
    of labor, See Wages
Cramer's rule, 106–07
Cummings, Ronald G., 47–48

Davidson, Paul, 268–69
Deep seabed mining, 87–88, 111–13,
    122–23

social costs, 127; theory of, 202−05; in relation to mining, 204−10

Extraction
alternative levels of, 38−40; cost of, 19−20, 32−33, 91−94, 234−35; holdbacks, 229−30; economic reasons for, 230−32; theory of, 230−33; marginal cost of, 100−02; of exhaustible resources, theory of, 47; optimal amount of, 38−39; optimal level of, affected by uncertainty, 40−41; optimal rate of, 41−42, 252−53; production function, 13−14; production process, 2; rate of, 15−17, 28−29, 32−33, 35−38, 40−42, 101−02, 111−12, 130−31, 272; time period of, 15, 231−32

Factor substitution
See Input substitutability
Fall, Albert B., 253−58
Free market
See Competition
Fuss, M.A., 65−66, 78, 80

Gapmanship, 89−90, 120, 122
Garrett process, 87−88
Gordon, Richard L., 42, 46−47
Government
constraints, See Environmental controls, Taxation; imposed conservation, 238−42; Teapot Dome affair, 253−60; imposed production holdbacks, 249−50, 252; intervention, 2−3, 28−30, 229; policies, 81−84, 134−35; cost to society, 238−40; price fixing, 248
Gray, Lewis C., 35−37
Greene, William H., 66
Gregory, Paul R., 66
Griffin, James M., 66

Harberger, Arnold C., 266−67, 272−73, 294−95
Herfindahl, Orris C., 42−46,
Homotheticity, 63−64, 66, 67, 69, 70−78; and technical change, 78−81; in nonmetal mining production function, 80−81
Hotelling, Harold, 36−38, 44
Hydrolic fracturing, 81−88, 91

Imports, effect of, 275−76
Income
discount factor, 154−56; from investment, 158−59; net, 154−55; See also Revenue
Inflation, 113−14, 191, 194−96, 225−26, 246, 258−59
Input
price of, 41−42, 67−69; substitutability, 64−65; empirical evidence, 65−66; usage, affected by costs, 142−43; usage, affected by environmental controls, taxation, 141−69; usage, Canadian mining industry, 66−69; usage, changes in, 73−74, 83−84; usage, in non-metal mining, 80−81; usage, in relation to output level, 63−64
Interest rate
and investment, 150−52, 165−68, 188−90; and marginal return to investment, 147−49, 163−65; changes in, 17−18, 46, 122−23, 147; compound, 232−34; effect on extraction rate, 36−37; effect on production holdback, 231−33; externally set, 48−49; in relation to prices, 46−49; in relation to resource value, 239−41; in relation to royalty, 44, 46; market, 146−47, 242−49; opportunity cost, 248−49; yield on, 247−48
Investment
affected by taxation changes, 183−88; affected by taxation, environmental controls, 141−69, 180−87; after-tax return rate, 188−90; and exploration, 20−24; and interest rate, 147, 150−52; capital, 147−49; cost of, 21−22, 151−52, 154−56, 157−58, 165−66, 179−80; decline in, 180−88, 188−91; depreciation of, 157−58; effect of environmental controls on, 145−50, 153−57, 161−69; effect of environmental controls without tax changes, 180−82; effect of taxation, environmental controls on, Canada, 176−300; effect of taxation on, 158−59; effect of taxation and environmental controls, 183, 187−88; expansion, 51−53, 163−65; expected returns, 21−22; expenditure, 160−62; in relation to changes in net worth, 166−67; in relation to percentage depletion, 266−72, 284, 286; in relation to resource price, 242−47, 262−63; in relation to return rate, 62; income from, 158−59; increase in, 266−73, 289, 292, 302; intramarginal,

allowance on, 266−67, 268−69; elasticity, 32−33, 167, 169, 268−69; equilibrium price, 96−117; expected, 32−33, 100, 108, 113−14, 230−32, 249−53; factor prices, linear homogeneity, 71−72; firm's effect on, 16; higher prices resulting from new technology, 89−91, 94−97; in relation to demand, 119; in relation to pollution expenditure, 136−37; in relation to supply, 27−28, 97, 99; in relation to total revenue, 91; influenced by supply and demand, 24−25, 28−29; market, 116−17, 122, 230−31, 234−36; of capital services, 67−69; of input, 41−42, 67−69, 93−94, 109−10; of non-conventional technology, 113, 114, 117−21; estimation, 114−17; of output, 41−42, 111−12; of resource, 113−14, 120−21, 234−37; change in, 102, 114−15; 119, 120−21, 234− 35; growth in, 242−49, 259−60; historical data, 242−49; in relation to AAA bond rates, 252−53; price behavior, 46−47; price effect of depletion, 284, 287−90; rise in, 46−48, 51, 249−53, 284, 294−97; to consumers, 266−67, 268−69; *See also* Royalty

Production

cost of, 27−28, 234−37; holdbacks, *See* Conservation, Stockpiling; relation to mineral extraction econometric model, 69−74

Production function, 13−14; cost function, 69−73; determining usage level of non-conventional technology, 120; empirical evidence, 63−84; fixed proportions, 117, 119, 120−21; for metal mining, empirical evidence, 73−81; for non-metal mining, 80−84; homotheticity of, 63−64, 67, 69, 70−78, 78−81; of extraction, 13−15; technology in, 69−70, 117

Profits

curve, 41; determining exploration rates, 43; discounted stream of, 15−16; "excess," 28−30, 32; incentive, 262−63; in relation to cost, 91, 93−94; maximization of, 2−3, 37−38, 134, 144−45; ratio to book value, 58−60, 274−79; related to exploration rate, 44, 46; to capital employed ratio, 53−57, 190−91, 194, 196−97; to equity ratio, 53−57, 190−91, 194, 196−97

Profitability, estimates of, 274−84

Property

area of, 43; non-owned or publicly owned, 130−32; owners, 205−07; rights, 132−33, 202−03

Rents, 30−32; definition of, 30; taxation of, 35−36

Reserves, petroleum

Teapot Dome Affair, 253−60

Resource deposit

multi-grade, 48−49; value of, 48

Return rate

after tax, 58−61, 115−17, 150−52; and environmental controls, 143, 145−47, 153; and investment, 151−52, 163−66, 178−79, 258−60, 269, 270−72, 274−84; before and after-tax, 52−53, 151−53, 157−59, 188−97; decreasing, 119−20; equalization of, 30; estimators of, 190−91; expected, 148−49; firms trading in U.S. financial markets, 54, 57−62; in Canadian mining industries, 52−57; in manufacturing before taxation, 242−48, 248−49, 252, 258−59; in relation to investment, 62; marginal, 104−05, 117; increases, 108; maximization of, 35−36; net, 153−54; on additional investment, 188−90; on capital, 50−53, 62, 114, 195−96; on AAA bonds, 246−48, 250−52, 259; ratio to capital turnover, 267−68

Revenue, 100; and investment, 154−55; reduction in, 143−45; net marginal, 104−05

Risk

See Uncertainty

Royalty, 30−32, 44, 46, 100, 257, 258−60, 268−69

Scott, Anthony T., 41−42

Severance tax, 46

Sinclair, Harry, 254

Smelting, 20−21

Smith, Vernon L., 47−48

Solow, Robert M., 48−49

Stockpiling, 228−63; as an investment, 249−52, 261; short-term, 249−53; theory of, 230−33

Substitutability of inputs, 25−26; degree of, 64−65, 69−73, 74, 77, 78, 81−83; empirical evidence, 65−66